Cheaper, Faster, Better?

Commercial
Approaches
to Weapons
Acquisition

Mark Lorell   Julia Lowell
Michael Kennedy   Hugh Levaux

Project AIR FORCE
RAND

Prepared for the United States Air Force

The research reported here was sponsored by the United States Air Force under Contract F49642-96-C-0001. Further information may be obtained from the Strategic Planning Division, Directorate of Plans, Hq USAF.

RAND is a nonprofit institution that helps improve policy and decisionmaking through research and analysis. RAND® is a registered trademark. RAND's publications do not necessarily reflect the opinions or policies of its research sponsors.

Cover photo courtesy of Lockheed Martin Tactical Aircraft Systems, Fort Worth, Texas. The photo shows the Lockheed Martin F-16 final assembly line in Fort Worth.

Published 2000 by RAND
1700 Main Street, P.O. Box 2138, Santa Monica, CA 90407-2138
1333 H St., N.W., Washington, D.C. 20005-4707
RAND URL: http://www.rand.org/
To order RAND documents or to obtain additional information, contact Distribution Services: Telephone: (310) 451-7002; Fax: (310) 451-6915; Internet: order@rand.org

PREFACE

This report discusses the issues surrounding more effective utilization of the civilian industrial base by the Department of Defense (DoD) and the U.S. Air Force. The first section of the report focuses on the "dual-use" nature of civilian and military technologies, and the potential for integrating the civilian and military industrial bases. The technology area studied, radar-related and other radio-frequency microwave devices, has traditionally been defense-specific. The second section of the report identifies mechanisms for minimizing the risks of inadequate product performance and excessively high cost in less-regulated commercial market environments. It discusses how well these mechanisms have worked in a defense-relevant commercial sector—the large transport aircraft industry—as well as in several experimental and pilot programs initiated by the Air Force and other defense agencies and services.

Most of the information and data for the material in this report were acquired through interviews with government and program managers and officials, and with industry officials. In addition to case studies and a wide array of published materials and other sources, the authors draw on a larger body of RAND research on the future of the defense industrial base.

This research was sponsored by the Office of the Assistant Secretary of the Air Force for Acquisition. It took place within the Resource Management and System Acquisition Program of RAND's Project AIR FORCE. It should be of interest to government and industry personnel concerned with weapon system acquisition, military acquisition reform, and avionics R&D strategies.

PROJECT AIR FORCE

Project AIR FORCE, a division of RAND, is the Air Force federally funded research and development center (FFRDC) for studies and analyses. It provides the Air Force with independent analyses of policy alternatives affecting the development, employment, combat readiness, and support of current and future aerospace forces. Research is performed in four programs: Aerospace Force Development; Manpower, Personnel, and Training; Resource Management; and Strategy and Doctrine.

Civil-military integration (CMI) lies at the core of current DoD efforts to reduce the costs of procuring and maintaining modern weapon systems. Results from this new and controversial approach are beginning to appear for several programs. This in-depth analysis, based on extensive interviews within government and industry, concludes that:

- Trends in DoD market shares for military-relevant avionics parts, together with supplier base shrinkage and obsolescence issues, indicate that the use of commercial-grade parts will increase dramatically whether or not a comprehensive CMI strategy is already in place. This conclusion is strengthened by the observation that many improvements in military avionics capabilities are being driven by the information revolution. Commercial firms are at the forefront of advances in information technology, making effective utilization of the civilian industrial base an important national security priority.

- Evidence from a case study of radio-frequency/microwave military avionics systems suggests that often commercial-grade parts, components, processes, and design elements can be successfully inserted without degrading system performance. Systematic insertion of commercial parts and technologies, combined with dual-use automated manufacturing processes, is also likely to reduce research and development (R&D) schedules and significantly reduce the costs of at least some types of military avionics modules. However, concerns remain about their long-term reliability and durability.

- A CMI strategy for military avionics cannot simply involve substitution of a few electronic components. Experience indicates that successful CMI may require that contractors be granted greater configuration control and change authority not only during the R&D phase but throughout the life-cycle of the weapon system. This raises the potential for a fundamental change in the role of the contractor and the current military depot system. More analysis is needed.

- Current DoD policy on procurement of "commercial items" does not reflect the true variety and complexity of commercial contract arrangements. What constitutes "commercial business practice" differs depending on the characteristics of the product and product market considered. In markets for complex, customized products, the most innovative commercial buyers and suppliers are increasingly establishing long-term partnerships (strategic supplier alliances) that involve extensive information sharing (including cost data) and flexible adjustment of price targets rather than strict firm-fixed-price contracts.

- Analysis of the commercial transport aircraft industry suggests a set of mechanisms relevant to the Department of Defense (DoD) that can lead to lower costs, reduced cycle times, and higher-quality items. These include the "must cost" approach to contracting, supplier base reduction, increased sharing of information and risk between buyers and suppliers, and greater outsourcing of life-cycle system support services to original equipment manufacturers. An important caveat remains the desirability of military programs to incorporate less mature, higher-risk technologies, which are largely avoided in the commercial market.

- Case studies of several military pilot programs that use key elements of the commercial approach reveal that a commercial-like acquisition strategy can be applied to military-unique items and that significant benefits, particularly in terms of production price, can be achieved. In three munitions programs examined that used CMI, the likely acquisition prices appear to be considerably less than half of what they would be in a traditional military procurement program, and operational performance capabilities appear to be meeting or exceeding original requirements. Most of the savings are the result of requirements reform, commercial parts insertion, and contractor ownership of and responsibility for design, technical content, performance, reliability, and price. These encouraging results indicate that similar reforms should be pursued aggressively in other DoD programs.

CONTENTS

FIGURES

TABLES

Declining defense budgets and growing weapon system procurement costs, combined with advances in commercial technologies, have led some observers to advocate the integration of the U.S. civilian and military industrial bases, a concept commonly referred to as Civil-Military Integration (CMI). As discussed in Chapter Two of this report, CMI advocates believe that DoD adoption of CMI would

- Reduce costs of acquiring and supporting weapon systems

- Improve performance throughout the life-cycle of a weapon system

- Shorten development times

- Improve reliability and maintainability

- Help maintain the defense-relevant portion of the industrial base.

To achieve the benefits of CMI, advocates call for a relaxation of the government-imposed oversight framework of regulatory restrictions that segregate weapon system acquisition from common commercial market practice and impose a regulatory cost premium on items purchased by the government. They base their arguments on two types of assumptions. First, they assume that an extensive "dual-use" overlap between commercial and military process and product technologies creates the potential for significant economies of scope and scale. Second, they assume that commercial business practices, together with the competitive incentives and constraints provided by a commercial-like market structure, will spur the development of high-performance weapon systems at lower cost than can be achieved under the current heavily regulated military acquisition process.

Critics of CMI respond that there is in fact little dual-use overlap between commercial and military products and processes in most crucial technology areas, so that integration of the defense and commercial industrial bases is simply not possible. Thus, CMI offers no real prospect for increased economies of scope or scale. Further, critics argue, DoD adoption of commercial business practices is unlikely to ensure DoD access to high-performance weapon systems at a reasonable price. They believe that relaxation of close government supervision over product development, together with removal of various other regulatory safeguards, will simply result in price-gouging by contractors and perhaps even outright fraud.

THE NATURE OF THE DUAL-USE OVERLAP: RF/MICROWAVE DEVICES

Trends in DoD market shares for many military-relevant avionics parts, together with supplier base shrinkage and obsolescence issues, indicate that the use of commercial-grade parts will increase dramatically whether or not a comprehensive CMI strategy is already in place. There is little question that, in some technology areas that are not defense-unique, the commercial sector is more technologically advanced than the military sector. These areas, sometimes referred to as the "low-hanging fruit" of DoD acquisition reform efforts, include items such as personal computers and certain microprocessors. Many efforts are now under way with little controversy to procure such items in the same way as in the commercial world.

For a rigorous comparison of the claims and counterclaims about the nature of the dual-use overlap between commercial and military product and process technologies, however, it is necessary to examine a product area that historically has been dominated by defense purchases. We select the area of radio frequency (RF)/microwave devices because it has traditionally been largely defense-specific, and because products such as fighter fire-control radars and fighter electronic warfare systems are of critical importance to future Air Force combat aircraft capabilities.

The questions we address are:

- Is the commercial market in military-relevant electronics large enough to encompass an adequate range of technologies, parts, and components required to support a comprehensive CMI strategy for military-specific microwave subsystems such as fire-control radars? (Chapter Three)

- Is the civilian market driving technology at a rate and in a direction that meets national security requirements? In other words, can CMI provide the necessary and desired performance capabilities? (Chapter Four)

- Are there cost and schedule benefits from inserting commercially derived parts and technology into RF/microwave systems? (Chapter Five)

We conclude that, in the critical area of defense-related microwave and millimeter-wave technologies, the dual-use promise of CMI is likely to be realized. Specifically, we find that the civilian market is beginning to drive the development of new technologies and lower-cost manufacturing processes in many types of RF/microwave products relevant to the military. Commercial design methodologies and process technologies are of growing relevance to military avionics system design and development, and commercial demand for many types of sophisticated RF/microwave parts and devices has already outstripped military demand. Commercially developed RF/microwave parts and components are also becoming increasingly available for incorporation in military systems. This implies that the technological breadth and depth in the commercial RF/microwave market necessary to support a comprehensive CMI strategy are emerging.

We also find evidence to suggest that commercial-grade parts and components can be successfully inserted into RF/microwave military avionics systems without degrading system performance. In fact, some commercially derived designs, technologies, and processes have the potential to increase performance. However, legitimate concerns remain about the long-term reliability and durability of commercial-grade parts and components.

Our response to the question of cost and schedule benefits is generally supportive of the dual-use nature of at least some RF/microwave technologies. The systematic insertion of commercial parts and technologies, combined with dual-use automated manufacturing processes, has the potential to shorten R&D schedules and reduce costs. There is some evidence to suggest that such a strategy could reduce costs of digital military avionics modules by as much as 20–50 percent. To achieve these cost and schedule benefits, however, contractors may have to be granted greater configuration control and change authority not only during the R&D phase but throughout the life-cycle of a weapon system. More analysis of this question needs to be undertaken.

ELEMENTS OF A COMMERCIAL-LIKE APPROACH TO ACQUISITION

Whether or not there is a large overlap between civilian and military products and processes, most CMI advocates believe DoD would benefit from adopting a more "commercial-like" approach to weapon system acquisition. Advocates claim that reliance on common commercial business practices, rather than regulation, will improve DoD access to low-cost, high-performance products.

But, as discussed in Chapter Two, there are many possible interpretations of what constitutes common commercial business practice, interpretations that depend on the nature of the products and product markets being considered. We therefore examine the commercial transport aircraft market, which has many structural similarities to the military aircraft market, to see what sorts of mechanisms commercial aircraft integrators, the suppliers of major aircraft systems, and the suppliers of aircraft parts and components are using to minimize their risks. We argue that these mechanisms may also effectively reduce the risks associated with the development, production, and maintenance of large and complicated weapons systems.

To assess the extent to which, and with what success, commercial-like approaches are being applied in military weapons programs, we examine experimental and pilot programs initiated by the Air Force, the Defense Advanced Research Projects Agency (DARPA), and the other services that contain key elements of the commercial approach. The programs we review include three "smart" munitions programs [the Joint Direct Attack Munition (JDAM), the Wind Corrected Munitions Dispenser (WCMD), and the Joint Air-to-Surface Stand-Off Missile (JASSM)]; two programs for the development and possible production of high-altitude endurance unpiloted air vehicles (DarkStar and Global Hawk);[1] plus an innovative modification program, DoD's Commercial Operations and Support Savings Initiative (COSSI).

The questions we address in Chapters Six and Seven are:

- What mechanisms have commercial market participants evolved to reduce risks associated with the development, production, and maintenance of large transport aircraft? To what extent are they relevant to DoD? (Chapter Six)

- To what extent, and with what success, have commercial-like approaches based on market mechanisms been applied to military programs, and what can be learned from them for future efforts? (Chapter Seven)

With respect to the first question, we find that the binding cost constraints introduced by airline deregulation have encouraged airframe integrators to forgo economically unnecessary performance innovations in order to adhere closely to production price targets. To achieve these targets, commercial airframe integrators increasingly are partnering with a small number of domestic and international suppliers with whom they share sensitive cost and technical data. These suppliers often provide their own financial capital as well as technical capabilities. To reduce risks for all concerned, the airframe integrators are strengthening lines of communication among their airline customers, their suppliers, and themselves through mechanisms such as integrated product teams (IPTs). Their contract structures are flexible, reflecting the relative commitments of the partners. For example, where risks are relatively low, such as for commodity purchases, firm-fixed-price contracts or simple purchase orders are common. Finally, to promote the reliability and maintainability of systems over time, there is a growing trend among air-

---

[1]The DarkStar program was cancelled in early 1999.

lines to outsource life-cycle system support services to original equipment manufacturers on the basis of fixed-price contracts.

Our examination of the commercial aircraft industry suggests that many commercial business practices are indeed relevant to military acquisition programs. Examples include IPTs, "best value" sourcing through preferred providers, and flexible contract structures that allow varying degrees of risk-sharing. Nevertheless, it can still be argued that important differences remain between the commercial aircraft and the military aircraft markets. Most notable of these are, first, the existence (in most cases) of a single buyer in military markets, and second, the tendency (or necessity) of military programs to incorporate high-risk (less mature) technologies to achieve the highest possible performance. These differences raise serious questions in the minds of some CMI opponents as to the applicability of commercial market mechanisms to military acquisition programs.

Our final set of case studies of the JDAM, JASSM, and WCMD munitions programs strongly indicates that a commercial-like acquisition strategy can be applied to military-unique items, and that significant benefits, particularly in terms of production price, can be achieved. In all three munitions programs, the likely acquisition prices appear to be considerably less than half of what they would be in a traditional military procurement program, and with the large procurement numbers involved, this should result in significant savings to the government. Further, although R&D is not complete on all of these programs and some technical problems have been encountered, operational performance capabilities appear on the whole to be meeting or exceeding original requirements, and the R&D phase for all three has remained relatively short by traditional military developmental standards.

In our judgment, the key aspects of the munitions pilot programs that have helped to achieve these successes are:

- Requirements reform and a closer customer-developer relationship through mechanisms such as IPTs, within an environment of binding cost constraints

- Contractor ownership of and responsibility for design, technical content, performance, reliability, and price.

However, we also believe that contractors could have taken on greater price risk in the production phase as well as cost risk in the R&D phase. Commercial transport aircraft developers price even their first aircraft according to a projected average recurring and nonrecurring cost over a relatively large production run, even though they have no guarantee that they will sell any aircraft at all. This is because customer airlines in the commercial marketplace would not tolerate paying the high price necessary to cover the actual recurring costs to the manufacturer of the early production aircraft. The experiences of DarkStar, Global Hawk, and COSSI suggest that greater cost risk sharing between the government and defense contractors is also possible, resulting in significant further cost reductions to the government of weapon system acquisition.

Finally, we conclude by cautioning that a CMI acquisition strategy does hold potential for a real loss of military capability. To the extent that the old flexible budget approach to acquisition resulted in weapon systems with many unnecessary features ("gold-plating"), a more commercial-like approach to weapon system acquisition would lead to a more responsible, cost-effective stewardship of the nations' resources. However, to avoid accidentally sacrificing features that may be crucial to successful mission completion, DoD must thoroughly understand the performance, schedule, and cost priorities for each new weapon system it solicits. Perhaps no less important, DoD must be able to communicate effectively those priorities to the program managers and private-sector contractors who have responsibility for making the tradeoffs between them.

ACKNOWLEDGMENTS

This study would not have been possible without the considerable assistance of numerous Department of Defense (DoD) and Air Force officials, defense contractors, and individuals connected to the commercial transport aircraft and electronics industries. We would especially like to acknowledge the many helpful comments and suggestions we received from Michael Vangel (Boeing McDonnell Aircraft and Missile Systems) and Richard Charles (Georgia State University). Invaluable comments were provided by the formal reviewers of this report: Christopher J. Bowie (Northrop Grumman Corporation) and Frank Camm of RAND. Camm's contribution to our analysis of commercial business practices and their applicability to DoD was particularly important. The authors of course remain responsible for any errors in fact or interpretation.

Finally, we would also like to acknowledge two RAND colleagues without whose continuing support we could not have completed this study: Natalie Crawford, RAND Vice President and Director of Project AIR FORCE (PAF), and Robert Roll, Director of the PAF Resource Management Program.

ABBREVIATIONS AND ACRONYMS

A/C	Aircraft
ACARS	Aircraft Communications Addressing and Reporting System
ACAT	Acquisition Category
ACC	Adaptive Cruise Control
ACTD	Advanced Concept Technology Demonstration
ACTS	Advanced Communications Technology Satellite
AFMC	Air Force Materiel Command
AIMS	Airplane Information Management System
ALCAR	Advanced Low Cost Aperture Radar
AOG	Aircraft on Ground
APU	Auxiliary Power Unit
ASG	Avionics and Surveillance Group (TRW)
ASIC	Application-Specific Integrated Circuit
AUPP	Average Unit Production Price
AUPPR	Average Unit Procurement Price Requirement
BFE	Buyer-Furnished Equipment
C&L	Coopers & Lybrand
C/SCS	Cost/Schedule Control System
CAB	Civil Aeronautics Board
CAIG	Cost Analysis Improvement Group (DoD)
CAIV	Cost As an Independent Variable
CAS	Cost Accounting Standards

CDRL	Contract Data Requirements List
CEP	Circular Error Probable
CICA	Competition in Contracting Act
CMI	Commercial-Military Integration
CNI	Communication, Navigation, and Identification
COSSI	Commercial Operations & Support Savings Initiative
COTS	Commercial Off-the-Shelf
CPFF	Cost Plus Fixed Fee
CPIF	Cost Plus Incentive Fee
CRADA	Cooperative Research and Development Agreement
CSIS	Center for Strategic and International Studies
CTS	Continuous Transverse Stub
DAPP	Defense Acquisition Pilot Programs
DARPA	Defense Advanced Research Projects Agency
DBS	Direct Broadcast Satellites
DCAA	Defense Contract Audit Agency
DCMC	Defense Contract Management Command
Dem/Val	Demonstration and Validation
DFARS	Defense Federal Acquisition Regulations Supplement
DMS	Diminishing Manufacturing Sources
DoD	Department of Defense
DRB	Dispute Resolution Board
EIA	Electronic Industries Association
EIWS	Electronic & Information Warfare Systems group (Northrop-Grumman)
EMD	Engineering and Manufacturing Development
ESSD	Electronic Sensors and Systems Division (Northrop-Grumman)
EW	Electronic Warfare
FAA	Federal Aviation Administration
FAR	Federal Acquisition Regulations

FARA	Federal Acquisition Reform Act (1996)
FASA	Federal Acquisition Streamlining Act (1994)
FFF	Form, Fit, and Function
FFP	Firm Fixed Price
FMS	Flight Management System
FY	Fiscal Year
GaAs	Gallium Arsenide
GAO	U.S. General Accounting Office
GDP	Gross Domestic Product
GHz	Gigahertz
GOTS	Government Off-the-Shelf
GPS	Global Positioning System
HAE	High-Altitude Endurance
HDTV	High Definition Television
HSCT	High Speed Commercial Transport
IAI	Israeli Aircraft Industries
IBP	Industrial Base Pilot
IIR	Imaging Infrared
IMU	Intertial Measurement Unit
IOC	Initial Operational Capability
IOT&E	Initial Operational Test and Evaluation
IPT	Integrated Product Team
IRS	Inertial Reference System
JASSM	Joint Air-to-Surface Stand-Off Missile
JDAM	Joint Direct Attack Munition
JSF	Joint Strike Fighter
JSOW	Joint Stand-Off Weapon
KPP	Key Performance Parameter
LAN	Local Area Network
LCR	Low Cost Radar
LEO	Low Earth Orbit
LRIP	Low Rate Initial Production

LRU	Line Replaceable Unit
LTBF	Least Time Between Failures
LTBUR	Least Time Between Unscheduled Removals
MAFET	Microwave and Analog Front-End Technology
Mantech	Manufacturing Technology
Mantech IBP	Manufacturing Technology Industrial Base Pilot
MCU	Modular Concept Unit
MDAP	Major Defense Acquisition Programs
MEO	Medium Earth Orbit
MHz	Megahertz
Mil-Specs	Military Specifications and Standards
Mil-Std	Military Standard
MIMIC	Microwave and Millimeter-Wave Integrated Circuit
MIRFS	Multifunction Integrated Radio Frequency System
MMIC	Monolithic Microwave Integrated Circuits
MMW	Millimeter Wave
MODAR	Modular Radar
MPM	Microwave Power Module
MRO	Maintenance, Repair, and Overhaul
MTBF	Mean Time Between Failures
MTBUR	Mean Time Between Unscheduled Removals
NASA	National Aeronautical and Space Administration
NAVSTAR	Navigation System Using Time and Ranging
NFF	No Fault Found
NRE	Non-Recurring Engineering
N mi	Nautical Mile
O&S	Operation and Support
OEM	Original Equipment Manufacturer
OH/M	Overhaul/Maintenance
OPP	Out of Production Parts
OSD	Office of the Secretary of Defense
OTA	Office of Technology Assessment

OTS	Off-the-Shelf
OUSD/A&T	Office of the Under Secretary of Defense for Acquisition and Technology
OUSD/AR	Office of the Under Secretary of Defense for Acquisition Reform
Pan Am	Pan American Airways
PB	President's Budget
PCS	Personal Communication Systems
PDR	Programmable Digital Radio
PDRR	Program Definition and Risk Reduction
PPCC	Production Price Commitment Curve
PPV	Past Performance Value
R&D	Research and Development
RF	Radio Frequency
RFP	Request for Proposals
RSAT	Radar System Aperture Technology
SATCOM	Satellite Communications
SFE	Supplier-Furnished Equipment
SIA	Semiconductor Industry Association
SLAM-ER	Stand-off Land Attack Missile-Expanded Response (Boeing)
SOO	Statement of Objectives
SOW	Statement of Work
SPO	System Program Office
SST	Supersonic Transport
T/R	Transmit/Receive
TBD	To Be Determined
TCAS	Traffic Alert and Collision Avoidance
TINA	Truth in Negotiations Act
TRP	Technology Reinvestment Program
TSPR	Total System Performance Responsibility
TSSAM	Tri-Service Standoff Attack Munition

TWA	Trans World Airlines
UAV	Unpiloted Air Vehicles
UPS	United Parcel Service
VARTM	Vacuum-Assisted Resin-Transfer Molding
VCR	Video Camera Recorder
VIA	Versatile Integrated Avionics
WCMD	Wind Corrected Munitions Dispenser
WLAN	Wireless Local Area Network

Chapter One

INTRODUCTION

OVERVIEW: THE PROMISE OF CIVIL-MILITARY INTEGRATION (CMI)

After World War II, the Pentagon's unique regulatory and oversight requirements, combined with its specialized and highly demanding military technology needs, led progressively to the emergence of two separate industrial bases: one for military research and development (R&D) and production, and one for the civilian market. Throughout much of the Cold War, the most advanced technology developments in many areas arose in the highly regulated military R&D sector. Beginning in the 1970s, however, civilian technology in electronics and other areas began overtaking and even surpassing developments in military R&D. In the 1980s, there was growing interest in developing strategies for more effectively exploiting commercially developed technologies for defense applications.[1]

By the mid-1990s, the problems of declining defense budgets and growing weapon system procurement costs combined with these technology trends to lead some officials in government and industry to advocate the integration of the U.S. military and civilian industrial bases, a concept commonly referred to as Civil-Military Integration

---

[1]The focus in this study is on DoD weapons acquisition as opposed to DoD acquisition in general because many of the most difficult issues surrounding integration of the civilian and military industrial bases arise from the military-unique nature of weapon systems. We recognize, however, that there is also significant cost-savings potential from greater integration of the military and commercial markets for non-weapons goods and services.

(CMI).[2] According to advocates, Department of Defense (DoD) adoption of CMI would

- Reduce costs of acquiring and supporting weapon systems

- Improve performance at Initial Operational Capability (IOC) and throughout the life cycle of a weapon system

- Shorten development times

- Improve reliability and maintainability

- Help maintain the defense-relevant portion of the industrial base.[3]

Some advocates have argued that the cost savings to DoD from CMI could be as high as $45 billion annually.[4]

The advocates of CMI base their arguments on two sorts of assumptions about the nature of the commercial versus military worlds. First, they assume there is an extensive "dual-use" overlap between commercial and military process and product technologies. According to advocates, many technologies, manufacturing processes, parts, components, and so forth are directly applicable to both commercial and military products, so that labor, research, production facilities, and other types of plant and equipment can be treated as dual-use. Second, they believe that DoD adoption of commercial business practices, in the context of the incentives and constraints provided by a more commercial-like market structure, will spur the development of high-performance weapon systems at lower costs than can be achieved under the current heavily regulated military acquisition process. Many advocates argue that, by adopting

---

[2]Although the term "CMI" takes on a variety of meanings in the literature, one commonly used definition is "the process of uniting the defense and commercial industrial bases so that common technologies, labor, equipment, material, and/or facilities can be used to meet both defense and commercial needs" (OTA, 1994, p. 42). We broaden this definition slightly to include DoD adoption of commercial market practices.

[3]The rationale for CMI and its expected benefits are discussed in several DoD documents, including Office of the Under Secretary of Defense for Acquisition and Technology (OUSD/A&T) (1995, 1996).

[4]See, for example, Gansler (1995).

commercial-like business practices, DoD can benefit from CMI even if militarily relevant products and technologies are not fully dual-use.

In basic agreement with the arguments of the CMI advocates, DoD leaders have begun to relax the web of regulatory restrictions that segregate weapon system acquisition from common commercial market practice and impose a cost premium on items purchased by the government. DoD leaders now insist that both the incentives and the constraints that prevail in military markets must be altered so that DoD and its defense contractors can begin to behave more like "normal" commercial buyers and sellers. This could require more extensive reform of the regulatory environment—and, in particular, reform of the Federal Acquisition Regulations (FAR) and Defense Federal Acquisition Regulations Supplement (DFARS)—than has already been achieved through ongoing DoD efforts. But probably more important, advocates say, to achieve the full benefits of CMI will require full and effective implementation of existing reforms by the military services.

Critics respond that even extensive acquisition reform will not result in the benefits promised by advocates of CMI. They argue that in fact there is little dual-use overlap between civilian and military products and processes in many crucial technology areas, so that integration of the defense and commercial industrial bases is simply not possible. They also believe that without regulatory safeguards, competitive incentives and constraints will be inadequate to ensure DoD access to high-performance weapon systems at a reasonable price. Given this market failure, they argue, a specialized cadre of defense-oriented firms operating under close governmental supervision is the best solution to U.S. national security needs.

Further, even those who believe in the basic promise of CMI have concerns about the effectiveness of its implementation by DoD. One worry is that Pentagon acquisition personnel may not receive adequate training and support to make the often-difficult decisions required of them in a commercial-like environment. For example, DoD managers may find it difficult to separate "must-have" system requirements from those that are only "nice-to-have," so that appropriate cost/performance tradeoffs are not made. DoD managers may also be reluctant to surrender control over weapon system configuration to contractors, but their failure to do so would reduce contractors' ability to make the changes necessary to provide many of the benefits promised by CMI.

RESEARCH OBJECTIVES

This monograph report explores the issues surrounding the more effective utilization of the civilian industrial base by DoD and the U.S. Air Force. We focus on two issues: the dual-use applicability of certain commercial technologies, and the risks and benefits to DoD of moving from a regulation-based approach to acquisition to a more commercial-like approach in which DoD emulates the practices developed by commercial customers. A brief history of how the U.S. civilian and defense industrial bases came to be separated is presented in Chapter Two, which also lays out the principal arguments and evidence for and against CMI.

In Chapters Three through Five, we discuss to what extent CMI might be appropriate to radio frequency (RF) microwave devices, a traditionally defense-specific technology that is of critical importance to future Air Force combat aircraft capabilities. Specifically, we seek to answer the following questions:

- Is the commercial market in military-relevant electronics large enough to encompass an adequate range of technologies, parts, and components required to support a comprehensive CMI strategy for military-specific microwave subsystems such as fire-control radars? (Chapter Three)

- Is the civilian market driving technology at a rate and in a direction that meets national security requirements? In other words, can CMI provide the necessary and desired performance capabilities? (Chapter Four)

- Are there cost and schedule benefits from inserting commercially derived parts and technology into military systems such as RF/microwave systems? (Chapter Five)

In Chapters Six and Seven, our research objective is first to identify business practices and strategies used in the commercial aircraft industry that could lead to a more cost-effective structuring of Air Force weapon system programs, and second to examine and assess experimental and pilot programs initiated by the Air Force, the

Defense Advanced Research Projects Agency (DARPA), and the other services that contain key elements of the commercial approach. The questions are:

- What mechanisms have commercial market participants evolved to reduce risks associated with the development, production, and maintenance of transport aircraft? To what extent are they relevant to DoD? (Chapter Six)

- To what extent, and with what success, have commercial-like approaches been applied to military programs, and what can be learned from them for future efforts? (Chapter Seven)

Our conclusions are summarized in Chapter Eight.

THE DUAL-USE APPLICABILITY OF COMMERCIAL TECHNOLOGIES

We begin by identifying candidate technology areas and a set of case studies to examine the dual-use applicability of commercial technologies. For this study, we select avionics because of its growing importance and cost for combat aircraft, and because the size and vitality of the nonmilitary electronics market should provide ample opportunities for CMI. To make the problem more manageable, we narrow our focus to defense-specific RF/microwave devices in the following two applications:

- Fighter fire-control radars

- Fighter electronic warfare (EW) systems.

Within these categories, we pay particular attention to the problem of developing cost-effective electronically scanned phased-array antennas. Here the key technological areas of interest are the transmit/receive (T/R) modules that populate the antenna array and that employ advanced gallium arsenide (GaAs) monolithic microwave integrated circuit (MMIC) chips.[5] Finally, we also examine some of the less exotic digital and RF devices used in fighter Communication,

---

[5]For an overview of the enabling technologies, see McQuiddy et al. (1991).

Navigation, and Identification (CNI) systems.[6] We chose these technological areas for three reasons: (1) the continuing growth of cost in fighter avionics; (2) indications that RF/microwave technology is opening up to increasing commercialization, thus offering CMI opportunities that did not exist in the past; and (3) high-level Defense Department advocacy of greater CMI in the field of military radars.[7]

We conducted case studies of programs involving a variety of military-specific RF/microwave technologies that, when taken together, represent many of the key dual-use elements of CMI. Some of them are pilot programs or innovative R&D efforts funded by the government and aimed at incorporating key aspects of CMI and other acquisition reform measures. Others are contractor-funded attempts to develop new low-cost systems based on commercial technologies. Some of the programs have produced only paper studies, but most involve the development of test hardware.

The following nine programs involve technologies directly applicable to RF/microwave, including fire-control radars and electronic warfare systems:

- Multifunction Integrated Radio Frequency System (MIRFS) Program

- Radar System Aperture Technology Program (RSAT)

- Advanced Low Cost Aperture Radar Program (ALCAR)

- Low Cost Radar Program (LCR)

- Modular Radar Program (MODAR)

- Modular Radars (AN/TPS-74)

---

[6]For the importance of microwave technology in a broad spectrum of defense applications, see Bierman (1991).

[7]For example, Dr. Paul Kaminski, former Under Secretary of Defense for Acquisition and Technology, stressed the significance of three key elements of a strategy for affordable radar systems: effective concepts of operations, compatible system architectures, and "the need to pursue acquisition approaches that leverage the broadest possible commercial industrial base." In a discussion on the future of radar technology in 1996, Dr. Kaminski pointed out that "one of the principal objectives of our acquisition reform program is to open the defense market to commercial companies and technology—not only the primes, but subtier suppliers as well." Quotations from Kaminski (1996).

- AN/ALQ-135 upgrade/support

- AN/ALQ-99 upgrade/support

- Technology Reinvestment Program (TRP): RF/Microwave/ Millimeter-wave (MMW) technologies.[8]

The first four are related to the development of new technology, lower-cost electronically scanned array (ESA) radar systems or antennas, and the next two involve development of low-cost conventional military radars. The two AN/ALQ case studies involve support and upgrade of existing EW systems. The last program is a government-sponsored effort aimed at developing microwave and millimeter-wave devices with both defense and commercial applications.

In CNI technologies, we examined the following four programs:

- Manufacturing Technology Industrial Base Pilot Program (Mantech IBP): Military Products from Commercial Lines

- Integrated Modular Avionics: F-16

- Integrated Modular Avionics from Commercial Lines: F-22

- Programmable Digital Radio (PDR).

The first three are related innovative fighter programs, and the fourth is a commercially developed component of a CNI system.

We chose these programs for our case studies because they are all characterized by one or more of the following attributes:

- Significant use of commercially derived designs or technologies during R&D

- Insertion of commercial components and parts

- Employment of dual-use production facilities or manufacturing technologies (mantech)

---

[8]The official program title is "Development and Application of Advanced Dual-Use Microwave Technologies for Wireless Communications and Sensors for IVHS Vehicles," but the scope of the effort has been broadened considerably since this title was formulated.

- "Spin-off" of military technologies to commercial applications with an eye to future "spin-back" of commercial technologies to defense applications.

We acquired most of the information and data on these programs through interviews with government program managers and other government officials, and with industry officials. Companies whose representatives were interviewed included:

- Raytheon Sensors and Electronic Systems, El Segundo, California[9]

- Northrop Grumman Electronic Sensors and Systems Sector, Baltimore, Maryland[10]

- TRW Space and Electronics Group (ASG), San Diego, California

- Northrop Grumman Electronic Systems, Rolling Meadows, Illinois

- M/A-COM, Lowell, Massachusetts

- AIL Systems, Inc., Deer Park, New York

- Raytheon Corporation, Washington, D.C.

- Northrop Grumman Xetron, Cincinnati, Ohio.

We also consulted a wide array of published materials and other sources. Of particular interest were materials provided either directly or indirectly by the following organizations:

- Avionics Directorate, Wright Laboratory, Air Force Materiel Command (AFMC), Wright Patterson Air Force Base

- Manufacturing Technology Directorate, Wright Laboratory, AFMC, Wright Patterson Air Force Base

- Electronic Systems Center, Hanscom Air Force Base

---

[9]In January 1997 Raytheon acquired the military components of Hughes and absorbed them into the Sensors and Electronic Systems Division of the Raytheon Systems Company.

[10]Formerly the primary military products division of Westinghouse Electric Corporation.

- Defense Advanced Research Projects Agency (DARPA)
- Electronics Industry Association (EIA)
- Semiconductor Industry Association (SIA).

A COMMERCIAL-LIKE APPROACH TO ACQUISITION

The second part of the report broadens the scope of our investigation to consider how DoD might benefit from relying on relevant commercial business practices rather than regulation to obtain low-cost, high-performance weapons systems—whether or not products and technologies are fully dual-use. The first of our two case studies summarizes the experience of a relevant commercial market; the second looks at DoD's own initial experience in a variety of ongoing pilot programs aimed at testing a commercial-like approach to acquisition.

For the first case study, we began by identifying commercial markets that share important structural features in common with the market for weapon systems. Using as selection criteria characteristics such as high technology content, large fixed costs of R&D and production, high performance and reliability requirements, and relatively small numbers of buyers and sellers, we chose to examine the commercial transport aircraft industry. We analyzed the arrangements that commercial airlines, airframe integrators, and aircraft equipment and parts suppliers have set up since the 1979 deregulation of the airline industry to control costs and ensure good performance over the life-cycle of an aircraft. We examined

- "Must cost" pricing structures and their implications for cost/performance tradeoffs
- Closer buyer-supplier relationships, including sharing of R&D, testing, and certification costs
- Buyer strategies such as cooperative acquisition, open industry standards, and "best value" sourcing
- Contractor configuration control and continuous technology insertion over the life-cycle of systems and subsystems.

Many of these arrangements are highly relevant to DoD acquisition of military aerospace systems.

We next identified existing military acquisition reform initiatives and pilot programs that attempt to approximate a commercial acquisition environment, and assessed these programs on the basis of design and implementation. Although all of the DoD acquisition pilot programs are limited in their application of commercial practices, they offer lessons to those considering a more widespread adoption of CMI. To capture these lessons most fully, three types of programs were considered:

- Service munitions programs that from their inception focus on developing military-unique combat systems under the direction of the user services

- DARPA acquisition technology demonstration programs that operate outside the normal acquisition environment

- Service modification and upgrade programs.

Our primary sources of data and information for both the commercial aircraft industry and DoD pilot program case studies were industry interviews and public materials available either in hardcopy or over the web. Companies whose representatives were interviewed included:

- Boeing Commercial Aircraft Group, Seattle, Washington

- Lockheed Martin Skunk Works, Palmdale, California

- Boeing McDonnell Aircraft and Missile Systems, St. Louis, Missouri[11]

- Boeing Douglas Products Division, Long Beach, California

- Northrop Grumman Integrated Systems and Aerostructures Sector, Dallas, Texas

---

[11]McDonnell Douglas was acquired by The Boeing Company in August 1997. Although some interviews were conducted after the acquisition, for simplicity we use the former McDonnell Douglas name.

- Northrop Grumman Integrated Systems and Aerostructures Sector, Hawthorne, California

- Raytheon Beech Aircraft Company, Wichita, Kansas

- Raytheon Sensors and Electronic Systems, El Segundo, California

- Teledyne Controls, Los Angeles, California.[12]

Only nonproprietary or generic information is presented throughout the report to permit wide distribution of our findings.

---

[12]Teledyne Controls is a fully owned subsidiary of Allegheny Teledyne Inc.

Chapter Two

ACQUISITION REFORM AND THE EVOLUTION OF THE U.S. WEAPONS MARKET

OVERVIEW: WEAPONS MARKET VS. COMMERCIAL MARKET STRUCTURE

In the United States, the market for defense-related goods and services is not like most commercial markets.[1] To understand the promise—and the risks—associated with CMI, it is useful to understand how the current military acquisition system and the separation between the defense and civilian industrial bases evolved.

The modern U.S. "market" for weapons and weapon systems has two dominating features. First, it is characterized by a single buyer, DoD, which defines the product and controls the sales opportunities of weapon system providers.[2] Unlike most commercial product markets, the weapons market is centralized, historically driven by the detailed performance and technical requirements provided by DoD.[3] Even where commercial markets for military technologies—or, per-

---

[1]There are, of course, many types of commercial markets, including markets for highly customized products. As we suggest in our case study of the commercial transport aircraft industry, these markets—not mass consumer product markets—are likely to provide the best model for a successful CMI strategy.

[2]Historically, the individual services have exercised considerable influence over acquisition decisions. As a result, one can plausibly argue that DoD is not monolithic, so that multiple buyers do exist in some circumstances.

[3]As will be seen later, a major thrust of acquisition reform is to reduce the level of detail in requirements and to move from technical specifications to broader performance requirements.

haps more likely, foreign military sales—may be possible, access to these markets is controlled by DoD and other U.S. government agencies. In contrast, for practically all mass consumer products, private firms have considerable control over both the R&D process and the configuration of the ultimate product. Sellers take the initiative in deciding what to produce, how much to spend on development, how to carry out R&D, how to test the finished product, and what price to charge. Diverse and autonomous buyers choose products offered by competitive sellers on the basis of their price and performance characteristics.

The second distinguishing feature of the weapons market is that it is characterized by a higher degree of technical complexity and innovation than most commercial product markets. To achieve DoD performance requirements, developers of new weapon systems not infrequently push the limits of known technology, incorporating both designs and materials that are largely unproven. Many if not most commercial product developers, in contrast, tend to improve incrementally on existing technologies. As a result, new commercial products do not usually differ in substantive technical ways from those already tried and tested in the marketplace.[4]

These two features of the weapons market imply two ways in which firms may find it riskier—and thus, without government support, less attractive—than commercial markets. First, the "market" risk associated with the weapons market may be higher. That is, defense contractors face a high risk that, after development and/or production costs have been incurred, the U.S. government will not choose to buy their product. Second, weapon system development and production may involve higher "technical" risk. That is, defense contractors face a high risk that the product will fail to achieve cost, performance, or delivery time objectives required by the U.S. government, and so fail to sell. Given that the expenditures on human as well as physical capital required for successful weapons development and production are substantial, the combination of these risks

---

[4]Exceptions, of course, exist. And while R&D expenditures as a percentage of sales in weapon systems are still much higher than for commercial products on average, the gap between commercial and defense-related expenditures on R&D is almost certainly less pronounced today than it was 30 years ago. See, for example, Peck and Scherer (1962).

may discourage firms from participating in the weapons business—
at least, not without a guarantee of an adequate rate of return on
investment.[5]

To ensure that the U.S. arsenal achieves high levels of technology,
the U.S. government has chosen to assume most of the risk of
weapon system development. DoD directly finances the bulk of the
R&D for most major weapons programs, its contracts are still primarily cost-based, and it tends to award sole-source production
contracts (effectively monopolies) to weapon system developers.[6] To
counter the potential for abuse of this sort of system, over time the
government has constructed an exceedingly complicated web of
acquisition-related laws, regulations, and practices that are incompatible with most standard commercial business practices. The U.S.
weapons market of the late 20th century is a far cry from the decentralized, full-information, price-based competitive market assumed
in simple economic theory—and assumed as well by some propo-
nents of CMI.[7] Further, past efforts at reforming the acquisition process have merely raised the barriers between the civilian and military
worlds.

ACQUISITION REFORM IN HISTORICAL CONTEXT

Recurrent problems with inadequate, underperforming, and overly
expensive weapon systems have led to calls for reform of the U.S.
military procurement system throughout its 200+ year history. Van
Opstal (1995) relates that one of the earliest attempts to revamp the
system occurred in 1777, when General George Washington was
forced to commission his own cannon-casting facilities because

---

[5]It is difficult to determine whether firms' stated reluctance to enter the defense
market arises from perceptions of excessive risk or excessive regulation. Gansler
(1995) claims that excessive regulation is responsible for most firms' unwillingness to
accept standard defense contracts with DoD; among the firms he cites are Hewlett
Packard and Digital Equipment Corporation.

[6]Of course, these monopolies are for specific weapon systems (the F-15, for example),
and other firms produce other weapon systems that are at least partially substitutable
for them (the F-16, for example).

[7]At least as early as 1962, the essentially nonmarket character of the weapons
acquisition process was recognized and carefully analyzed by economists. Peck and
Scherer (1962) remains one of the best analyses of the U.S. weapons market.

private manufacturers were unwilling to accept the contract—an early example of the problems created by perceptions of excessive market risk. During the United States' first 100 years or so, however, there were few changes to U.S. military technology, so that technical risk was low. Inventories were small and remained in service for long periods,[8] making possible a highly decentralized weapons procurement process, with little formal coordination between the services. The individual technical branches of the services took "an approach to technological advance that emphasized the strict separation of R&D from production, elaborate test procedures, competitive bidding for defense contracts, and quality control during production" (McNaugher, 1989, p. 21).[9] Although not a recipe for handling either the pressures of wartime or rapid technological change, it did serve to minimize the potential for misuse of the taxpayers' money.

The inadequacies of this approach became painfully evident at the outbreak of World War I, especially with respect to rapidly evolving aircraft technologies. In the time it took to write a detailed fixed-price contract and conduct a competition on its terms, the technology had often become obsolete.[10] For aircraft, the problem was temporarily solved by circumventing the services' formal procurement procedures altogether and turning to the private sector for both design and manufacture. Private companies took on the major responsibility for the technical management and integration of military aircraft programs, with only rather loose direction provided by formal military contracts.

By the end of World War I, political problems with reliance on the private sector for military aircraft development had begun to surface.

---

[8]For example, Holley (1953) describes in great detail the U.S. Army's reluctance to convert from muzzle-loading to breech-loading rifles. Although patented breechloaders existed as early as 1842, they were not widely available as standard issue until 1865—just *after* the end of the Civil War.

[9]For the decentralization of the procurement process prior to World War I, see Peck and Scherer (1962).

[10]Holley (1953) points out that the original 225-horsepower design for the Liberty aircraft engine, which was recommended for production in April 1917, became obsolete just three months later. By the end of World War I, the Liberty engine had been transformed from 8 cylinders to 12 cylinders, and from a 225- to a 440-horsepower rating.

In response to reports of wartime profiteering, Congress demanded that all military design work once again be competitively awarded using detailed fixed-price contracts. Further, Congress insisted that the services acquire technical data rights from the winner of the original design competition to put production contracts out for competitive bid. The net result was that some aircraft and engine firms refused to bid on contracts that gave the government proprietary rights to the finished design. Nevertheless, Congress refused to reconsider its requirement for competitive bidding on military contracts, even though technical progress for military aircraft was considered unsatisfactorily slow.[11]

This situation continued until after the European outbreak of World War II in 1939, when Congress signed emergency legislation granting the services wide latitude to negotiate contracts, including cost-plus, sole-source contracts "if the emergency demanded" (McNaugher, 1989, p. 29). After the war, Congress gave primary responsibility for military acquisition to the newly created Department of Defense, but it was the also newly created U.S. Air Force that drove the budget and to a large extent determined acquisition policy.[12] For the next forty-odd years the Air Force, and DoD as a whole, continued to rely primarily upon the cost-plus, often sole-source weapons contracting system that had proven so effective during the war. The system, however, became increasingly burdened by regulations designed to prevent the smallest violation of the public trust. Over time there evolved an almost complete separation between commercial markets and the military acquisition process, from the most sophisticated weapon systems to the smallest parts and components. Where it has been possible to compare the two, the achievements of commercial markets often appear to have been superior in terms of product cost, timeliness, and sometimes even performance.

---

[11]McNaugher (1989, p. 28) argues that throughout the decade of the 1920s, private contractors were forced to absorb an increasing share of the risks associated with military aircraft design and production without sufficient offsetting compensation. As a result, "many aircraft firms lost interest in military business and shifted instead to the commercial side of their operation."

[12]Peck and Scherer (1962) suggest there may have been doubts about the Secretary of Defense's authority over the procurement process as late as 1958. During the latter half of the 1950s, the Air Force accounted for 47 percent of the overall defense budget, with spending on aerospace R&D and production rising throughout the decade (McNaugher, 1989).

The tension between DoD's desire to uphold the public trust and the need for fast, effective provision of military systems has continued throughout the Cold War period and beyond. As a result, efforts to reform the acquisition process have tended to split along conceptual lines: those designed to fix problems attributable to self-interested and even criminal behavior on the part of public officials and defense contractors, and those designed to fix problems attributable to inflexibilities imposed by governmental regulations. These two types of "reform" have worked mostly at cross-purposes. For example, in the immediate pre- and post-Sputnik years, many would-be reformers argued that substantial improvements would result if accepted business practices were to replace government regulation as a guide to U.S. weapons acquisition (Peck and Scherer, 1962).[13] On the other hand, perceived contractor "waste, fraud, and abuse" led to the passage of the Truth in Negotiations Act (TINA) in 1962, the creation of the Defense Contract Audit Agency in 1965, and the establishment of Cost Accounting Standards (CAS) in 1970. In the 1980s, the President's Blue Ribbon Commission on Defense Management (the Packard Commission) recommended that DoD expand its use of commercial products and institute "commercial-style" competition on the basis of past performance as well as price. This recommendation, however, was at least partially offset by other reforms such as the 1984 Competition in Contracting Act (CICA), which sought to ensure equal access to defense contracts for all firms regardless of size or experience.

RECENT EFFORTS TO REFORM THE DEFENSE ACQUISITION PROCESS

Beginning in the late 1980s, as pressure to reduce the federal budget deficit began to mount, an increasing number of observers both inside and outside the Pentagon concluded that the pendulum had swung too far in the direction of regulatory oversight, creating an increasingly unnecessary separation between the civilian and defense industrial bases. For example, in a series of influential books and ar-

---

[13]Interestingly, both Peck and Scherer (1962) and Scherer (1964) concluded that the adoption of "commercial practices" would do little to improve the weapons acquisition process because of what they considered to be the fundamentally different characteristics of the weapons market, as discussed above.

ticles, Under Secretary of Defense for Acquisition and Technology Jacques Gansler—who was at that time working in the private sector—argued that the heavily regulated military acquisition system not only discouraged efficient defense-related production but actually failed the public trust by encouraging defense contractors to produce unnecessary and unnecessarily expensive items.[14] Gansler and other reformers identified two serious and related problems with the U.S. defense acquisition process.

First, the reformers pointed out that many commercial firms were consciously avoiding Pentagon business because of government-mandated procedures and standards that were not in conformity with routine business practices in the commercial world. Those firms that did work on DoD contracts tended to either specialize in military work or establish separate divisions fenced off from their commercial divisions so that government regulations and oversight would not impinge on their commercial operations. Thus, reformers argued, the maze of government-unique requirements and standards acted as a barrier to DoD acquisition of relatively inexpensive yet state-of-the-art commercial product and process technologies. They asserted that this problem was especially acute in information technology, which has the potential to radically increase military effectiveness even without increases in weapon system platform performance, but which is almost entirely driven by developments in the commercial sector.

Second, the reformers argued that firms' compliance with the various laws and regulations related to government procurement, combined with the extra cost of mandated government monitoring and oversight, caused a significant cost premium to be added to items procured by the government. According to studies conducted in the late 1980s and early 1990s, government regulation increases costs to the government for various weapons programs by 5 to 50 percent.[15]

---

[14]See, for example, Gansler (1989, 1995).

[15]Some of the leading studies of that period, along with their estimates of the DoD cost premium, are Smith et al. (1988), 5–10 percent; OTA (1989, Vol. II, appendix), 10–50 percent; CSIS (1991), 30 percent; Carnegie Commission on Science, Technology and Government (1992), 40 percent; American Defense Preparedness Association (1992), 30–50 percent.

The solution to these problems, according to reformers, was for DoD to encourage greater integration of the defense and civilian industrial bases. Appropriate steps include dismantling the regulatory and informational barriers to the use of dual-use process and product technologies, and replacing those regulations with appropriate commercial business practices designed to keep costs down and product quality and performance up.

In response to these and other criticisms and suggestions, Congress passed Section 800 of the National Defense Authorization Act of 1990. This Act required DoD to establish a panel of experts from government, industry, and academia to evaluate changes to DoD acquisition regulations. Consistent with the CMI reformers' recommendations, the Section 800 Panel proposed eliminating or changing about one half of the 600 statutes it identified as affecting DoD acquisition. Its findings were submitted to Congress in January 1993 for legislative action.[16]

The findings of the Section 800 Panel, together with the work of Vice President Gore's National Performance Review, convinced top DoD leaders of the need for rigorous reform of the defense acquisition process and influenced their strategy for achieving it. In particular, Secretary Perry's February 1994 vision statement, *Acquisition Reform: A Mandate for Change*, called for a flexible, commercial-like approach to defense acquisition emphasizing increased use of commercial products, technologies, and processes and greater integration of the civilian and military industrial bases. Many of these ideas were subsequently incorporated in the Federal Acquisition Streamlining Act of 1994 (FASA), which greatly simplified DoD procedures for purchasing relatively low-cost, low-risk commercial products and services. Among other things, FASA

1. Expanded the definition of commercial items

2. Automatically exempted the purchase of commercial items from more than 30 government-unique statutes

3. Removed the requirement for cost and price data on commercial contracts

---

[16]See U.S. Congress (19 March 1997).

4. Raised the threshold for the application of TINA to $500,000

5. Expanded the information provided to all competitors after contract awards to reduce formal protests.

FASA also authorized the establishment of Defense Acquisition Pilot Programs (DAPPs), enabling the services and defense agencies to test out the more radical modes of acquisition reform.[17]

1994 also saw the initiation of another key component of DoD's acquisition reform policy—requirements reform. The first element of requirements reform, Military Standards and Specifications (Mil-Spec) reform, was motivated by the argument that the wholesale application of Mil-Specs to military programs was inhibiting the incorporation of advanced commercial technologies and processes into military products. Mil-Specs were also believed to discourage commercial firms that used only commercial specifications and standards from participating in military acquisition programs. To remedy this perceived problem, in June 1994 Secretary Perry issued a memorandum entitled *Specifications and Standards—A New Way of Doing Business*. This memorandum turned existing DoD policy on its head: Instead of requiring Mil-Specs, as had been the case in past policy, it called for the use of commercial and performance standards wherever possible, and required defense programs to provide special justifications if Mil-Specs were used.

A central aspect of Mil-Spec reform is that the service buyer should not dictate specific or detailed technical and design solutions to contractors. Instead, contractors should be provided with general system and performance requirements necessary to accomplish the military mission. As in the commercial world, defense contractors should be given more opportunity to develop new and innovative design and technical solutions at lower cost in order to meet the mission requirements.

Cost is also central to the second element of requirements reform, "Cost As an Independent Variable," or CAIV. CAIV requires DoD ac-

---

[17]The Federal Acquisition Reform Act of 1996 (FARA) made additional changes to promote even greater government access to the commercial marketplace, by further simplifying procedures for purchasing certain categories of commercial items. See OUSD/A&T (1996, Appendix B).

quisition managers to raise cost considerations to a priority level at least equal to, and often even higher than, the traditional military program requirements relating to system performance and development schedule.[18] CAIV is intended to mimic conditions in the commercial world, where cost is always an independent variable. It has two primary features.

First, CAIV requires that the government buyer—the services and DoD—have a clear and precise understanding of the mission for the system and what system outcome is needed on the battlefield. The buyer then must carefully prioritize the mission performance needs and broad capability requirements that the system should possess to accomplish the mission.[19] Prioritization is critical so that intelligent tradeoffs can be made between cost and capability. A principal objective of this approach is to avoid "gold-plating" weapon systems with extensive capabilities that are not truly necessary to perform the mission, but that often drive up costs by necessitating the use of unique military-only parts and technologies.

Second, for the CAIV process to achieve its full potential, reformers believe that contractor configuration control is necessary, at least below the overall system level. Configuration control combined with Mil-Spec reform permits the contractor to seek out and experiment with any technologies and parts available in the marketplace, whether commercial or military, in order to meet the government buyer's mission requirements at the lowest possible cost.

In sum, FASA, Mil-Spec reform, CAIV, and the other acquisition reform initiatives that have proliferated over the past ten or so years reflect the CMI advocates' belief that the civilian industrial base can

---

[18]As defined by Noel Longuemare, the Principal Deputy Under Secretary of Defense for Acquisition and Technology, "CAIV means that we will intentionally hold cost constant and accept the schedule and performance that results—within limits of course." Quoted in OUSD/AR (May 1996). The basic concept of CAIV is not dramatically different from the "Design to Cost" concept applied with mixed results in the early 1970s. The difference, according to advocates, is that CAIV is being implemented in an environment of much more profound change to the traditional acquisition system and culture. They believe that this gives it a much greater chance of success.

[19]These may include factors such as reliability, sortie rate, survivability, and robustness, along with more traditional measures of performance such as speed, range, and payload.

provide DoD with relatively inexpensive yet high-performance product and process technologies suitable for defense applications. Such measures also promise to reduce or eliminate the cost premium associated with military-unique regulations and standards. But why should these reform efforts succeed when so many before them have failed? Advocates have two answers: The growth of the dual-use technology sector and the effectiveness of commercial-world mechanisms for minimizing risk.

The Growth of the Dual-Use Sector

CMI advocates believe that technological developments in both the military and (especially) commercial worlds mean that process and/or product technologies used for commercial and defense-related design and manufacture are now similar, or in some cases identical. This means that the same people, machines, and facilities can be shared between defense and commercial applications. CMI thus not only offers economies of scale and scope in dual-use development and production, but also effectively reduces the market risk of weapons production by allowing firms to recover their fixed costs from many more potential buyers. In theory, this should make the weapons business more attractive to more firms, reducing DoD's need to pay for all of weapon system R&D.

But is there sufficient dual-use overlap to achieve the effective integration of the defense and commercial industrial bases? Much depends, first, on the nature of recent developments in commercial technologies, and second, on the complexity and requirements of the system under consideration. It is true that developments in commercial technologies, and to a lesser extent in defense technologies, have made certain commercial and defense applications more similar than was the case 15, 20, or even 40 years ago. But it may also be true that CMI is still relevant only to a subset of systems that DoD buys and yet can still achieve many of the benefits claimed for it by its supporters.

Although there are substantial overlaps between them, DoD purchases can be divided conceptually into three categories:

- pure commercial

- commercial but substantively modified for military use ("commercial-modified")

- military unique.

Pure commercial items include items the military buys that are identical to those bought in commercial markets. Examples include food, office space, clothing, gasoline, office furniture, medical care and supplies, and so forth. A few of these types of items are incorporated into weapon systems (some commercially available microchips, for instance), but even the parts and components of weapon systems generally require some type of modification.[20] The high costs to the military of pure commercial items relative to what they sell for in commercial markets can be attributed to burdensome government regulations and oversight, so the argument that CMI can save DoD money with respect to pure commercial items has been accepted by both DoD and Congress for some time. Increased use of pure commercial items became a formal element of DoD policy as early as 1976, and in 1984 Congress mandated DoD's procurement of pure commercial items "whenever such use is technically acceptable and cost-effective" (OTA, 1994, p. 64).[21] Significant steps have already been taken to rapidly integrate the defense and commercial markets for many of these items, and we do not consider them here.[22]

Of more current interest are items that have clear similarities to those produced in commercial markets but are modified for military use. Examples include most computers, global positioning system (GPS) receivers, space launchers, utility helicopters, and transport aircraft and trucks. Military modifications often involve ruggediza-

---

[20]Nondevelopmental items (NDIs), DoD's purchase of which has been strongly encouraged by Congress, include pure commercial items as well as previously developed military items, with allowance for some modification. The distinction between a modification and a redesign is the subject of much debate, and is discussed at greater length in Chapters Three through Five below.

[21]In 1993, DoD reported that commercial items accounted for approximately 7 percent of the funds spent on "high dollar value items." The estimated commercial shares of total procurement of parts relevant to weapons systems, reported by the Defense Logistics Agency (DLA) and the Air Force, were approximately 18 and 10–15 percent, respectively (DoD, 1993).

[22]Much less integration has occurred in the area of services.

tion so that the equipment will survive in combat environments, but generally do not involve major technical challenges. Items in this category may be alleged to cost more than commercially available alternatives, but it is difficult to separate those costs attributable to government regulations and oversight from those attributable to the customization. Because commercial markets for similar types of products already exist, however, it is probably here that CMI's greatest potential lies. Many if not most weapon system parts and components potentially belong in the commercial-modified category.

The most problematic area for CMI supporters lies in the category of military-unique items, items that, at least in the past, have had no close commercial analog. These items are designed and developed for military purposes, and include most weapon systems and many weapon subsystems. Examples include combat aircraft (especially stealthy supersonic ones), fire-control radar, guided missiles, nuclear weapons, and nuclear submarines. To some extent, military-unique items can be differentiated from commercial-modified items by their level of technical difficulty and complexity: Unlike commercial-modified items, military-unique items generally involve a technically challenging development process. Of more direct relevance to CMI, military-unique items

1. have no obvious commercial counterpart

2. largely use noncommercial processes

3. involve highly classified and controlled technologies.[23]

At least in the past, they have had little obvious potential for dual-use application.

Recent developments in commercial technologies are blurring the line between the commercial-modified and military-unique categories of items that DoD buys. The technological superiority of military relative to commercial technologies that was widespread in the 1950s, 1960s, and even 1970s is no longer so clear in the 1990s. The post-war paradigm of "spin-off" is turning into "spin-on": More and more, defense technologies are driven by developments in the com-

---

[23]This definition is taken from OTA (1994, p. 139).

mercial world. In particular, commercial developments in information technology can potentially increase DoD's ability to find, fix, and locate targets, allow rapid transmission of data, and make munitions more autonomous and precise. Thus statements (1) and (2) above are increasingly suspect, especially with respect to subsystems, parts, and components.

Early Evidence of the Benefits of Dual-Use

To date, little formal evidence has been provided to prove or disprove the assumption that many if not most military products and processes have become dual-use. Although a few studies have made careful empirical analyses of the potential gains to DoD from particular dual-use technologies (for example, OUSD/A&T, October 1996), the more general claims are still based on collections of anecdotes.

Probably the most familiar argument presented in favor of CMI is that commercially developed and produced items ("commercial items" for short) cost less than their military counterparts. Where such items are identical, or nearly identical, the military should be able to take advantage of commercial economies of scale by buying commercial off-the-shelf (COTS) items. Even where they are not, the dual-use aspects of design and manufacturing process technologies may make it possible to achieve economies of scope through modification of commercial items. This point is made by CMI advocates in a widely cited anecdote about the price of military computer chips. Reportedly, in recent years the military has paid ten dollars apiece for computer chips similar to ones costing just a dollar on commercial markets.[24] Another example is secure telephony. According to the Defense Science Board (1993), DoD's commercially derived STU-III secure telephone costs about one-tenth as much as a conventional Mil-Spec item and took Motorola just three years to develop as compared to an average 7–11 year DoD cycle. From this type of evidence, the Defense Science Board estimated that DoD procurement of

---

[24]The reference may be to the CMOS (Complementary Metal Oxide Semiconductor) chip developed commercially in Japan for wristwatch batteries and now widely used in military applications. See National Economic Council et al. (1995) and Alic et al. (1992, p. 73).

commercial items in electronics, software, and spare parts could result in savings of 3–20 percent.

A second and potentially even more important argument, given the reliance of U.S. national security strategy on qualitatively superior military systems, is that the commercial sector is technologically ahead of the military sector in those areas where both use broadly similar technologies. Once again, microchips are a widely cited example. The Packard Commission concluded in 1986 that "military microchips typically lag a generation (three to five years) behind commercial microchips."[25] Thus, it is argued that a newly acquired commercial computer today embodies more recent and thus more powerful technology than a newly acquired computer designed and developed for the military. This is generally attributed to the greater flexibility of commercial markets in incorporating technology into new products, compared with the time-consuming, costly, and generally burdensome process that military developers (and government monitoring authorities) must go through to get new technologies approved for government purchase.

CMI advocates also claim that, under the current acquisition process, the technology in legacy weapon systems tends to be frozen for long periods between occasional major upgrades. By contrast, many commercial systems have their technological components upgraded throughout system life, a process called "continuous insertion" of new technology. For example, Gansler (1995, p. 136) states that the mobile electric power unit used in many weapon systems is a 25-year old design that is less efficient, more polluting, and less reliable than available commercial units. More dramatically, it appears that certain parts for the yet-to-be-produced F-22 are already out of production.[26]

According to CMI advocates, the reasons why military systems tend not to have new technology continuously inserted include rigid adherence to outdated Mil-Specs, reluctance to surrender configuration control to contractors, and a lack of incentives for military R&D budget allocators to invest in upgrades. Investments in upgrades are

---

[25]As cited in Alic et al. (1992, p. 153).

[26]See the discussion in Chapter Five.

unpopular because the savings in operating costs cannot be retained by the military to enhance other capabilities the way they can in private firms.[27] It is argued, however, that military programs will soon be forced to adopt continuous insertion because of the lack of availability of original parts and components. We discuss this further in Chapter Five.

A third potential benefit claimed by CMI advocates is that commercial products usually have much shorter development cycles than do military systems. Commercial markets, spurred by competition, are said to develop new products more efficiently, so that new generations of products appear approximately every five years as opposed to every 15 for military systems. This may be partly because commercial development cycles are more incremental, and thus naturally more frequent, than are military development cycles. In contrast, in military development practice, a few new products, each of which incorporates major technical change, are introduced at infrequent intervals. "Block upgrades" to military systems are analogous to the commercial world's "new generation" of products, however, so this argument should be viewed with caution.

Furthermore, it is claimed that some commercial industries achieve significant schedule reductions—and avoid extremely costly redesigns—through close integration of the design, engineering, and production phases of the manufacturing process. For example, by requiring manufacturing and design engineers to work together at an early stage in the development process—as well as by extensive use of computer-aided design—the Boeing team responsible for the passenger doors on the 777 achieved close to a 95-percent reduction in manufacturing design errors (Sabbagh, 1996, p. 91). A similar lesson can be drawn from the McDonnell-Douglas TAV-8B, which had 68 percent fewer drawing changes and 58 percent scrap reduction as a result of integrating design and manufacture (Gansler, 1995, p. 184). As a result of these kinds of arguments and anecdotes, all new military acquisition programs incorporate the "Integrated Product Team" (IPT) approach.

---

[27]It is sometimes argued that the small fleet size of some military systems deployed also makes it difficult to amortize technical upgrade investments.

A final argument made by CMI advocates is that buyers of products in commercial markets do not face the sorts of "industrial base" problems that the military does. Because the military is often the most important customer for its supplier firms, temporary reductions in military purchases can cause supplier firms to fail. When this happens, it can leave the military with few or even no suppliers. If the military were part of a diversified commercial customer base, according to CMI advocates, a hiatus in military orders would pose less of a problem for its suppliers. Like glass or rubber manufacturers facing a downturn in the auto industry, military suppliers could weather the storm by turning to alternative product lines. Further, the danger of creating a supplier monopoly would be much less, because there are far fewer barriers to entry into the commercial market than into the current highly regulated defense market. Finally, with respect to an issue of considerable concern to military planners, CMI supporters claim that an integrated civil-military industrial base would provide the necessary flexibility to support a wartime surge situation. Just as commercial customers with urgent needs can pay for priority deliveries, in a surge situation the military could offer a premium for products to guarantee their delivery.[28]

Risk-Minimization and Commercial Business Practices

If the first assumption of CMI advocates is that many commercial product and process technologies are effectively dual-use, their second crucial assumption is that commercial business practices can be an effective replacement for government regulations. They believe that commercial market mechanisms for minimizing both technical and market risk will also keep costs down and quality and performance up in the context of defense acquisition. Their argument is that, while it still may be necessary to produce some (or even many) military items on purely military production lines, DoD could still benefit from taking a less bureaucratic approach to weapon system acquisition. In particular, some reformers have argued that DoD must become a "world-class" customer—and its suppliers "world-

---

[28]That this practice might be labeled war profiteering may make it a less-attractive strategy for the military *and* for industry.

class" suppliers—by adopting business practices characteristic of the very best commercial firms.[29]

Unfortunately, advocates as well as critics of CMI tend to differ in what they mean by "commercial business practices." There is therefore considerable confusion about what exactly DoD ought to do to achieve the benefits from such practices. We identify four major interpretations of commercial practice:

- Traditional—DoD's formal definition of all activities that can be provided by a nongovernment source as "commercial activities." Under this interpretation, defense contractor behavior that occurs only because a highly regulated approach to government contracting allows, encourages, or requires it is sometimes attributed to the commercial world in general.

- Textbook—the introductory economics textbook definition of commercial practice, in which firms rely on arm's-length competition based on firm-fixed-price (FFP) contracts to exchange products.[30] Such practices are most commonly found in markets involving generic goods and services, traded broadly and deeply, in which little specific investment or customization on either side of a transaction is required.

- Best—the commercial practices that characterize the firms recognized by their peers as being the best-managed firms in the world. A key element of this definition of commercial practice is flexibility, with the nature of business relationships and contract types adjusted to the complexity of the particular market. For complex, customized products, for example, such firms tend to emphasize strategic partnership rather than arm's-length contracting, with an emphasis on benchmarking, reputation building, and information exchange between buyer and seller.

- Official—commercial practice as defined in recent federal legislation, DoD acquisition reform initiatives, and changes in the FAR, particularly the introduction of FAR Part 12. This definition emphasizes expanded use of firm-fixed prices and best-value

---

[29]See, for example, Perry (1994).

[30]In the simplest textbook models, no contracts are involved; all transactions take place in spot markets.

competitions, as well as increased contractor management and control over design configuration and commercial-style warranties. A key element here is the idea that FFP contract structures allow commercial firms to stop collecting cost data from their suppliers, thus apparently eliminating a primary contributor to the regulatory cost premium.

Of these four interpretations of commercial practice, the "traditional" interpretation is clearly least relevant to CMI advocates—if not necessarily to their critics. Whatever it is that reformers wish to see, it is not the practices that defense contractors have developed over time in response to a highly regulated acquisition system, and we will not consider this interpretation further here. The "textbook" and "best" interpretations, however—each of which contains elements that are formally incorporated in the "official" interpretation—are both legitimate points of departure for devising DoD acquisition policies. Unfortunately, these two interpretations of commercial practice can have quite different implications for policy. Practices that may work well for highly liquid markets involving generic goods and services are much less likely to work well for markets in which products are highly customized and technical risks are high.

An example is pricing and the structure of supplier contracts. For those "textbook" commercial product markets in which contracts are common, FFP structures tend to dominate, for two reasons. First, because these markets feature products that are well defined and relatively homogeneous, with many possible buyers and sellers, relevant price information is readily available. FFP contracts work best when buyers and sellers can easily agree on appropriate price targets. Second, since both market and technical risks are minimal in this type of market, sellers are generally willing to cover cost overruns—because they have more information about and technological control over product development and production than do buyers.

On the other hand, in commercial product markets where technical and/or market risks are high and little price information is available, a multiplicity of fixed-price-type and cost-type contracts exists, allowing for various degrees of risk-sharing between buyers and sellers. In some markets, cost-plus-incentive-fee contracts similar to those still prevalent in DoD weapon system development programs are common, requiring the buyer rather than the seller to pay for most if not all unexpected cost increases.[31] In other markets, partnering relationships substitute for arm's-length contracts, with buyer and seller working together to achieve mutually agreed-on price, quality, and performance targets. Within such arrangements, both risks and returns are often shared equally, with failure to continue the relationship as the ultimate penalty for missing desired targets.

Research suggests that the "best" commercial firms pay close attention to the characteristics of both products and markets.[32] Instead of a blanket insistence on particular contract forms, these firms flexibly adjust their sourcing strategies according to their perceptions of both the risk and value of the product concerned. Their arrangements with suppliers run the gamut from purchase order and credit card arrangements, to price-based and cost-based contracts, to long-term partnering agreements. For products deemed to be very high value as well as very high risk, a lack of relevant data may preclude contract types requiring the ability to determine prices based on market- or model-based information. To better manage their costs in these situations, the "best" firms often choose to establish close corporate relationships with a limited number of suppliers. Buyer and seller agree to share sensitive cost, technology, and resource data and to greatly reduce the degree of competition relevant to each of them. Mutual commitment—the symmetry of the deal—is important to the persistence of trust, and so key to the success of the relationship.

In sum, CMI advocates agree that commercial business practices can help keep costs down and product quality and performance up, whether or not the economies of scope and scale associated with dual-use production are achievable. However, differences in interpretation of "commercial practice" imply quite different policy rec-

---

[31]In general, FFP contracts require suppliers to bear 100 percent of cost overruns, in contrast to cost-plus-fixed-fee–type contracts, which require buyers to bear 100 percent of cost overruns. In between these two polar cases lie a range of contract types with various arrangements for sharing risk. See, for example, the discussion in Rogerson (1992, pp. 11–12).

[32]See, for example, the undated corporate documents provided by John Deere, a commercial firm widely regarded for its innovative and effective supply management and purchasing policies. Recent RAND studies that explore DoD-relevant aspects of "best" commercial practices include Camm (1996) and Pint and Baldwin (1997). A non-RAND exploration of this topic can be found in Tang (1999).

ommendations. For example, CMI advocates who take a "textbook" view of the commercial world tend to assume there are a variety of price analysis techniques capable of revealing fair and reasonable prices for products of similar quality. They therefore advocate DoD adoption of FFP contract structures to help ensure that DoD obtains military items at those prices. In contrast, CMI advocates who want DoD to become a "world class" customer tend to encourage adoption of commercial practices that are "best" in the required context. They believe DoD can successfully establish relationships with some contractors that are based on mutual trust and benefit. They do not believe in the need for blanket adoption of FFP-type contracts to ensure protection against profiteering by unscrupulous suppliers.

EVIDENCE OF THE BENEFITS OF DEREGULATION

As mentioned above, several studies in the late 1980s and early 1990s indicated the existence of a large cost premium associated with the regulations governing defense acquisition. None of these studies rigorously quantified the premium, however, and none offered more than qualitative suggestions as to the biggest regulatory cost drivers. From a cost-benefit standpoint, there was no hard evidence to support decisions for or against throwing out either particular regulations or particular categories of regulations.

To get at this evidence, in 1994 former Secretary of Defense William Perry tasked a private consulting firm, Coopers & Lybrand (C&L), to undertake a detailed analysis of industry compliance costs. C&L joined with TASC, a systems engineering group with expertise in government procurement issues, to collect data at ten defense contractor sites.[33] Focusing on 130 DoD regulations and standards identified by the Section 800 Panel and others as major cost drivers,

---

[33]TASC became a fully owned subsidiary of Litton Industries in 1998. The ten contractors were Allison Transmission (General Motors), Beech Aircraft (Raytheon), Boeing Defense & Space Group, Rockwell Collins Avionics and Communications Division, Hughes Space and Communications Company (General Motors), Motorola Systems Solutions Group, Oshkosh Truck–Chassis Division, The Timken Company, Teledyne Ryan Continental Aviation Engine group, and Texas Instruments Defense Systems & Electronics Group (Raytheon). Some of these companies have since merged or been acquired by other entities; the new parent companies are in parentheses.

C&L/TASC concluded that on average DoD paid a regulatory cost premium of 18 percent.[34] The top ten cost drivers were found to account for about half of this cost premium; the top 24 accounted for 75 percent.

The official C&L/TASC findings were reported to DoD in December 1994 (C&L/TASC, 1994). In a series of follow-up studies, most of the DoD regulations and standards identified by C&L/TASC as driving up contractor compliance costs were also identified as major impediments to greater participation of commercial firms in DoD procurement, including dual-use procurement. In addition to commercial firms' apparent unwillingness to accept the extra costs associated with defense regulations, they were apparently reluctant to accept the governmental controls on profits and governmental access to proprietary technical and cost data required by DoD contracts.

The biggest contributors to the regulatory cost premium as well as the highest regulatory barriers to commercial participation appeared to fall within the following categories:

- Government access to commercially sensitive product cost and pricing data such as required by TINA

- Government-imposed accounting and reporting standards and systems such as CAS, Cost/Schedule Control System (C/SCS), and Material Management Accounting System

- Audit and oversight requirements such as Defense Contract Management Command program reviews, Defense Contract Audit Agency audits, and Contractor Purchasing System reviews

- Complex contract requirements and Statements of Work (SOWs)

- Mandatory socioeconomic source requirements

- Government ownership and control of technical data.

Most of these relate to contract structures or accounting and oversight procedures.

---

[34]In several subsequent studies, estimates of the regulatory cost premium are much smaller. See Lorell (forthcoming) for a discussion of these estimates.

But C&L/TASC and the others were not asked to compare or quantify the possible effects of adopting alternative commercial-like approaches to the regulations they identified as budget-busters. As a result, there was no attempt to show, for example, whether fixed-price–type contracts—with no cost reporting requirements—would be better able than more traditional cost-type contract structures to control problems such as program cost overruns and performance shortfalls. Nor was there any attempt to analyze whether DoD would benefit from broadening its competitions to include more nontraditional suppliers or from restricting them to a select group of "preferred suppliers." Any of these arrangements is possible; all exist in the commercial world.

Further, while C&L/TASC and other groups studying the regulatory cost premium found large potential cost savings from deregulation of defense acquisition, they all conducted what economists call "partial equilibrium" analyses. That is, they assumed that the factors affecting cost elements such as materials costs and contractor profits would remain unchanged if government regulations were eliminated. The implications of deregulation for competition and market structure were ignored. Perhaps more important, their studies do not address the question of how deregulation might affect the quality or performance of military items. As discussed below, many critics of CMI believe regulation is essential for DoD to maintain qualitative superiority in any future war.

THE CRITICS' RESPONSE

CMI critics are skeptical of both the dual-use and deregulation claims put forward by advocates. With respect to dual-use technology, they argue that items such as computer chips and secure telephones are special cases that cannot be extrapolated to the broader military acquisition environment. In general, they do not believe DoD can become just one more customer within a large and diversified customer base. They believe that differences between the kinds of products required by military as opposed to commercial customers are inherently unbridgeable. With respect to deregulation, they reason that DoD's unique mission requirements and substantial political constraints make "commercial business practices," whether interpreted as "textbook" or "best," unsuitable for DoD. Finally, some critics find that some aspects of CMI may be beneficial in theory, but they doubt the effectiveness of CMI implementation in practice either by DoD or by private-sector defense contractors.

Critics identify four sets of factors likely to cause CMI-based reforms to fail: factors affecting cost; factors affecting performance; factors affecting DoD implementation of CMI reforms; and factors affecting private-sector implementation of reforms.

Factors affecting cost. Factors identified by CMI critics as potential contributors to excessive cost under commercial-like acquisition programs include:

- Insufficient competition for DoD contracts

- Limited non-DoD sales opportunities for military items

- Parts proliferation resulting from elimination of Mil-Specs and increased contractor configuration management.

The possibility that a more commercial-like approach to acquisition will create opportunities for "excessive" contractor profits is a concern expressed by CMI critics. It is certainly true that, if DoD eliminates current restrictions, some contractors could earn much higher profits. In purely budgetary terms, these contractor profits should be irrelevant to DoD as long as they reflect declining costs rather than rising prices. But in political terms, large contractor profits could be difficult to explain to Congress and the tax-paying public. CMI critics worry that the elimination of DoD's profit policy could put the military services back in the situation they faced at the end of World War I, when Congress responded strongly and negatively to reports of wartime profiteering.

At some level, the key to this problem is competition: With sufficient competition, neither prices for weapons systems nor contractor profits will be "excessive."[35] But critics point out that DoD's current policy of paying for 100 percent of weapon system R&D effectively limits competition because the government cannot afford to pay the

---

[35]That is, defense contractors will earn a rate of return on investment that is comparable to returns earned in other competitive industries.

R&D for very many firms.[36] To increase the level of competition for DoD production contracts, therefore, one strategy would be to encourage contractors to finance a greater share of military R&D.[37]

Unfortunately, a second concern critics raise about a commercial-like acquisition strategy is that firms will refuse to do business with DoD unless either R&D cost recovery is guaranteed or the winners of production competitions are heavily compensated for their investment risk. The argument here is that, because military products and processes are not dual-use, DoD effectively acts as a single buyer. The replacement of 100-percent cost-plus R&D contracts with contracts incorporating greater degrees of risk-sharing will therefore prove unacceptable to firms because DoD cannot commit itself to future purchases. The financial incentives necessary to get firms to participate in weapon system contract competitions, CMI critics argue, will effectively wipe out any cost savings achieved by eliminating 100-percent cost recovery on military R&D. The more limited the potential for non-DoD sales, the higher the risk for which contractors must be compensated.[38]

Finally, a longer-term concern is that a commercial-like approach in which contractors are responsible for configuration management and control will result in parts proliferation. Specifically, critics argue that form-fit-function-test integration guidelines described in requests for proposals (RFPs) may be unable to prevent parts from becoming increasingly unique, so that the support and maintenance of weapon systems will become more and more expensive over time. In fact, some critics argue that parts proliferation could raise support costs even in the short term because small-lot production tends to lower production efficiency, increase maintenance training requirements, and increase record-keeping burdens. The counter to this argument is that the effect of diminishing manufacturing sources

---

[36]The problem of insufficient competition becomes particularly pronounced at the engineering and manufacturing development (EMD) and production phases.

[37]A related concern is that if firms overrun costs under a fixed-price contract, DoD will be compelled to accept a price increase as political forces mobilize to prevent major financial losses to the company (i.e., a bailout).

[38]Non-DoD sales could include both foreign military sales and commercial sales associated with dual-use products and technologies.

(DMS) on Mil-Spec parts and components even more severely affects short- and long-term support costs.[39]

Factors affecting performance. According to CMI critics, factors that may contribute to inadequate weapon system performance under commercial-like acquisition programs include:

- Large differences between commercial versus military usage and/or environments

- Commercial quality assurance and testing practices that are too tolerant of product variability

- Insufficient government oversight of programs

- Too great an emphasis on system cost as opposed to performance.

Critics believe that all four of these factors may be exacerbated if private-sector contractors are allowed to manage and control the configuration of weapon systems.

Perhaps the primary concern about the commercial approach to weapon system acquisition is that it could result in systems that cannot perform their intended missions. In particular, CMI critics worry that the insertion of off-the-shelf commercial products—or even nondevelopmental items based on commercial designs—into military systems could cause those systems to fail in military environments. If the environment and usage for which commercial items are designed are much less demanding than their expected military environment and usage, performance problems are likely to occur.

Critics worry that the adoption of certain commercial quality assurance and testing practices may also cause performance problems, particularly for military applications that require very low product variability. Because of the large output volumes involved in mass commercial manufacturing, quality assurance in these types of settings tends to involve probabilistic approaches such as simulated reliability prediction models. Although much cheaper than DoD-style individual product inspections, these practices are not as thorough.

_____

[39]See the discussion in Chapter Five.

Therefore, for weapon system parts and components that have low error tolerances, probabilistic testing practices may not be appropriate. In response, advocates point out that total quality management (TQM) places primary emphasis on eliminating variability in production through total process control. TQM is now standard practice among the "best" commercial firms, and it is not clear that the traditional DoD approach to quality assurance works better.[40]

According to critics, unsatisfactory weapon system performance may occur more often in a commercial-like acquisition environment where, once qualified, there is comparatively little oversight of suppliers.[41] Suppliers may misunderstand DoD's requirements in the absence of Mil-Specs, or they may even attempt deliberate fraud. In the case of deliberate fraud, most commercial firms operate on the premise that extreme vigilance is more costly than the fraud deterred. In a military environment, however, the smallest degree of fraud may be unacceptable, not only because of the enormous human and military consequences of military equipment failure, but also because of the negative political repercussions from revelations of fraud, which may put future congressional appropriations for weapons programs in doubt.[42]

Finally, a somewhat more subtle concern about the commercial approach to weapon system acquisition has to do with the nature of commercial-world tradeoffs between cost and performance. It is not yet clear whether commercial approaches with their heavy emphasis

---

[40]As pointed out by one of our reviewers, TQM and DoD quality assurance systems typically have different goals for reducing variability. DoD seeks to reduce variability to ensure future performance in demanding operating environments. The primary goal of most commercial TQM procedures is to reduce total ownership cost. Nevertheless, our reviewer argues that TQM programs can be developed that will reduce variability in any attributes of any part or system that DoD cares about.

[41]Oversight of suppliers can be far reaching in some commercial industries, but still tends to be less than that demanded by the U.S. government. In contrast to government, commercial buyers generally put a great deal of effort into choosing their suppliers. They then presume that the suppliers they have chosen will fulfill their contracts to the best of their ability. See, for example, the discussion of Boeing's relations with its suppliers in Sabbagh (1996).

[42]There are significant exceptions to this characterization of the "commercial world," including the civil aviation industry, the nuclear power industry, and various others. The importance of performance and safety requirements for civil aircraft is one reason why we chose the large transport aircraft industry for our case study in Chapter Six.

on reducing total ownership cost can produce the highly innovative, extremely high-performance technologies embodied in U.S. weapon systems such as jet fighters. The issue here is not so much failure to perform but rather failure to excel at warfighting. In certain military situations, having the best warfighting equipment, rather than merely good warfighting equipment, can make the difference between victory and defeat. In certain key weapon system programs, an overemphasis on cost could result in systems that are good but not good enough, thereby nullifying any cost savings achieved by adoption of a commercial-like acquisition approach. Another way of saying this is that, when implementing a CAIV approach to weapon system acquisition, DoD acquisition managers must realize that, in the commercial world, modest decreases in technical performance generally lead to only modest decreases in utility. In the military world, on the other hand, modest decreases in technical performance may lead to large decreases in military utility, with serious consequences for war-winning capability.[43]

Factors affecting implementation by DoD. CMI critics identify the several factors that may make it difficult for DoD acquisition managers to transition to a commercial-like acquisition strategy, including:

- Unclear statement of mission requirements by DoD leadership leading to poor understanding of performance, schedule, and cost priorities by acquisition managers

- Lack of familiarity with new and existing commercial technologies and standards and practices, and inadequate training to become familiar

---

[43]In a world of perfect certainty, in which the tradeoff between cost and performance were known exactly, this would not be a problem. In the certainty case, the processes of minimizing cost for given performance, maximizing performance for given cost, or balancing cost and performance to get the most cost-effective weapon system would be equivalent. The problem is that, in a world of uncertainty, the eventual outcome of a development program can be crucially affected by the emphasis placed on the different goals of the program. A program in which developers are urged to worry about cost will likely result in a lower-cost, lower-performance product than one in which developers are urged to worry about performance. If modest changes in performance lead to major changes in utility, commercial approaches to acquisition, which tend to have a strong emphasis on reducing cost, may be problematic.

- Inadequate mechanisms for DoD acquisition managers to communicate preferences and priorities to contractors

- Poor managerial incentives for DoD acquisition personnel.

A commercial approach to weapon system acquisition would put considerable responsibility on the shoulders of DoD program managers and acquisition personnel. In contrast to the system based on Mil-Specs, for example, the commercial approach gives contractors latitude to design a variety of engineering solutions to a particular performance requirement. Therefore, instead of merely verifying that proposals meet Mil-Specs and are the lowest bid, in a more commercial-like setting acquisition personnel must be able to judge both the cost-performance value of the contract and the adequacy and plausibility of particular technical solutions.

These judgments require that DoD managers have a thorough understanding of each system's intended missions at an early stage in the process in order to prioritize and communicate them to contractors. This is particularly true given DoD's renewed emphasis on CAIV. Performance requirements must be widely disseminated to ensure the participation of firms that could provide potentially superior solutions.

But the tradeoffs between various performance requirements are many-dimensional, and their implications for design and engineering decisions may be difficult to determine. Commercial technologies and standards may not translate easily to military applications. At least initially, a severe handicap for acquisition managers will be that, for any particular program, the probability distribution of outcomes based on commercial solutions is unknown. Further, there may be requirements embedded in existing Mil-Specs that do not get effectively included in performance requirements documents, because of a lack of institutional memory or simple oversight. Critics believe that, despite the best efforts of acquisition personnel and contractors, the selected solutions may not be optimal.

The risk of choosing the wrong technical solution increases greatly if DoD is not able or willing to bear the costs of adequate training. For example, if DoD personnel do not participate in the formulation and maintenance of international commercial standards, they may not know how to interpret or apply them in evaluating proposals. If DoD is unwilling to pay the upfront costs of participation in mechanisms for communicating its preferences, such as IPTs, it increases the risk that such communication will be inadequate. In addition, a lack of personnel trained to devise requirements tradeoffs, prepare RFPs, and select sources is at minimum likely to delay desired weapon system purchases.

Finally, critics believe that one of the most difficult problems facing DoD is how to create incentives that will encourage acquisition managers to support the new commercial approach wholeheartedly. Under the old system, managers were rarely rewarded if they chose low-cost solutions or solutions unfamiliar to DoD. Further, the adversarial acquisition environment taught DoD managers to distrust contractors. Learned behaviors such as these may be difficult to unlearn. Functional specialists such as contract managers, who have invested heavily in the old system, may require new training programs at considerable resource cost.[44]

Factors affecting private-sector implementation. Factors that may make it difficult for private contractors to transition to a commercial-like acquisition strategy include

- High and unrecoverable transition costs

- Resistance to increased competition and fear of losing specialized advantages

- Distrust of the political process.

The implementation risks associated with the private-sector transition to a commercial approach may also be considerable. Established defense contractors may be unwilling to spend the resources necessary to make a successful transition, particularly if they believe the costs of transition to be unrecoverable. For example, the

---

[44]As pointed out by a reviewer, these sorts of implementation-related concerns are probably always present in situations involving large institutional changes. The issue is therefore, not whether change is necessary, but rather whether DoD has properly prepared for it. Policy questions include

- What needs to change to make the new method work?

- How can we know when it is ready?

- Can we test it incrementally to limit the risk of moving into it?

transition to process quality control assurance, activity-based accounting methods, and full participation with DoD personnel in standards groups will require considerable private investment. Industry may also be resistant to participation in arrangements such as IPTs and Cooperative Research and Development Agreements (CRADAs) because they fear loss of trade secrets or other competitive advantages.

In fact, many private military contractors may have good reasons to prefer the old acquisition system to a commercial approach. Established contractors have experience in dealing with the old Mil-Spec system and have invested in large bureaucracies trained in the intricacies of government contract procedures. Why should they level the playing field and give new entrants equal access to the weapons business? Further, after so many years in a heavily regulated environment, defense contractors may not wish to incur the "organizational stress" of a fundamental change in the way they operate. In a commercial environment, their own lack of experience might put them at a disadvantage relative to aggressive newcomers.[45]

Finally, there may be historical reasons for contractors to prefer the protection of an arm's-length relationship with DoD. As mentioned above, in the 1920s and again in the 1960s and 1970s there was a strong political backlash against the perceived "cozy" relationship between DoD and its major defense contractors. It is entirely possible that commercial-world practices such as source selection based on "best value" criteria could once again lead to the anti-contractor political animus of earlier periods.

CONCLUSION

Efforts to reform the U.S. military acquisition system are almost as old as the system itself. As indicated in our brief historical review, many if not most past efforts have either failed to achieve their intended objectives or have achieved their objectives (such as discour-

---

[45]Again, some of the implementation risks outlined above are inherent to any transition process. To the extent that commercial firms have had to face similar sorts of restructuring in the past, they may have already developed mechanisms to deal with them.

aging fraud) at the expense of themselves introducing new problems (such as costly oversight regulations) into the system.

But the new group of acquisition reformers, who base their reform strategies on CMI, believe that their efforts will succeed where previous efforts have not. Their arguments rely heavily on the assumptions that, first, there is a large dual-use technology base waiting to be tapped by DoD, and second, that commercial business practices can effectively replace the current superstructure of acquisition regulations.

So far, however, there is little systematic empirical support for or evidence against these assumptions. In the chapters that follow we help to fill this gap with a series of relevant case studies.

Chapter Three

COMMERCIAL TECHNOLOGY TRENDS RELEVANT TO MILITARY RADARS

INTRODUCTION

A fundamental assumption made by the advocates of CMI is that commercially derived technologies, products, and processes increasingly are at least equivalent and often superior to those developed in the military sector. Skeptics, however, accuse CMI advocates of "cherry picking" a few obvious technology areas such as integrated circuits or microprocessors where great advances have been made in the commercial sector over the last few decades, and ignoring many other crucial technology areas where the military sector remains far ahead. Clarification of this question is critical. CMI cannot be viewed as a desirable strategy that will bring significant benefit if advanced technologies directly relevant to military applications do not exist in the commercial sector.

A definitive all-inclusive assessment of this issue is far beyond the scope of this research. Instead, we examine commercial market trends in a critical military technology area that traditionally has been dominated by military R&D and that is rarely used as an example by CMI advocates: defense-related microwave technologies. We seek to shed light on the following question:

- Is the commercial market in military-relevant microwave electronics large enough to encompass an adequate range of technologies, parts, and components required to support a compre-

45

hensive CMI strategy for military-specific subsystems such as fighter fire-control radars?

Questions on the performance capabilities and potential cost and schedule benefits of commercial microwave electronics technologies are explored in Chapters Four and Five.

RADAR AND THE NEW COMMERCIAL MARKET IN MICROWAVE TECHNOLOGIES

The development and deployment of radar is one of the great historical achievements of the military industrial base. Although the first practical use of radar can be attributed to American physicists conducting scientific experiments in 1925, most of the major technical and engineering innovations that made the widespread use of radar possible were developed by the military R&D establishments of the United States, the United Kingdom, and Germany before and during World War II. After the war, radar began to be used in many civilian applications, including weather avoidance, navigation, and maritime surveillance. Later, radar was used for high-resolution area mapping and for many civilian space applications. Nonetheless, the major technology developments in radar continued to be driven by the demanding performance and environmental requirements of military systems. This was particularly true in the 1950s and 1960s with the introduction of extremely sophisticated multirole fighter fire-control radars for air-to-air and air-to-ground operations.

Until recently, the vast majority of radio-frequency consumer products operated well below the 1 Gigahertz (GHz) frequency range on the electromagnetic spectrum. Fire-control radar, however, typically operated much higher up the electromagnetic spectrum in the X-band (8–12.5 GHz) and lower Ku-band (12.5–18 GHz) frequency ranges, thus requiring substantially different—and more demanding—hardware and technical and manufacturing techniques for parts and components.[1] Throughout most of the Cold War era, therefore, military radar and other military electronics requirements drove most of the technology developments in the microwave fre-

---

[1]For the development of radar in the 1930s and 1940s, see Buderi (1996) and Stimson (1983).

quency range (about 1 GHz to 30 GHz) and millimeter-wave (MMW) frequency range (30–100 GHz).

Phased-array radars, based on electronically scanning antennas populated with transmit/receive (T/R) modules that employ GaAs MMIC chips, are on the cutting edge of military radar technology. They provide numerous advantages over conventional radars, particularly for fighter aircraft, including lower radar cross-section (greater stealthiness), simultaneous multiple-target engagement capabilities, extended target-detection range, higher survivability, greater reliability, and reduced weight and size. All the original T/R module and electronically scanned array technologies were developed by military contractors using government money.

By 1990, however, a technology revolution appeared to be under way in the commercial sector regarding microwave and MMW technologies. As the decade of the 1990s comes to a close, the fundamental assumption of CMI advocates for our technology area of inquiry seems to be increasingly true: Many defense-critical RF microwave/MMW technologies directly relevant to military radars, CNI, EW, intelligence gathering, and other sensors appear increasingly likely to be driven by civilian market demands. If this is true, then military systems developers must efficiently exploit technology developments in the commercial sector to gain access to the most advanced technologies available.

There are four particularly active product areas in the emerging commercial microwave market relevant to military microwave products that exhibit great technological dynamism:

- Land-based wireless communications

- Television Direct Broadcast Satellites (DBS), and High Definition TV (HDTV)

- Automotive sensors

- Mobile communications satellite systems.

Their relative positions on the electromagnetic spectrum are shown in Figure 3.1.

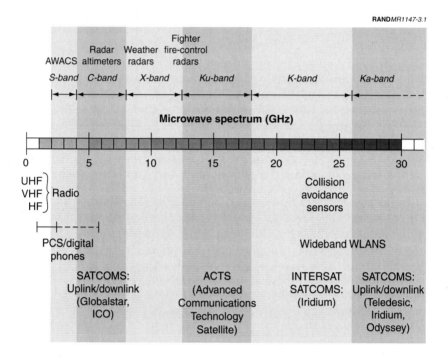

Figure 3.1—The Dual-Use Microwave Spectrum

Two highly active technology areas in land-based wireless communications are (1) cellular phones, cellular/fax-data modems, personal communication systems (PCS), and RF identification sensor systems, and (2) wireless local-area networks (WLANs). Automotive sensors include radar T/R modules being developed for vehicle collision avoidance or adaptive cruise control (ACC) systems. The new generation low-earth orbit (LEO) and medium-earth orbit (MEO) mobile communications satellite antennas and transceiver systems are extremely important for the commercial development of low-cost, high-capability T/R modules. Finally, the tuners and antenna receivers used in television DBS and HDTV are prominent in the commercial microwave revolution. Rapid market growth is projected for many of these defense-relevant RF/microwave commercial products.

WIRELESS COMMUNICATIONS

Dramatic growth is expected to continue in the wireless communications sector. High growth in the demand for cellular phones and PCS is expected to drive this increase. As shown in Figure 3.2, worldwide cellular phone subscribers are projected to rise from under 90 million in 1995 to almost 700 million in 2003. At the end of 1997, the worldwide digital subscriber base outnumbered the analog subscriber base for the first time. By 2003, digital technologies are expected to account for over 91 percent of the market.[2]

Whereas conventional cellular phones employ analog technology and operate in the 800 MHz frequency range and below, PCS use digital technology and operate at higher frequencies in the 1.8 GHz range and above. Higher-frequency broadband-width digital technology permits the transmission of far more information on the same

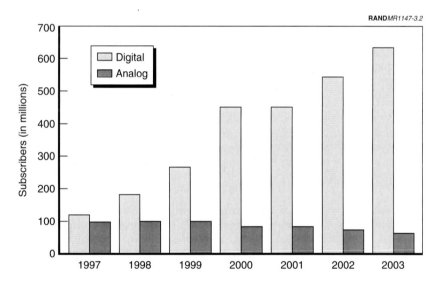

Figure 3.2—Projected Worldwide Cellular/PCS Subscribers by Technology

---

[2] Data from the Strategis Group, Washington, D.C.

channel. Thus PCS carriers will be able to offer many services in addition to the standard voice transmission available with current analog cellular phones, including wireless access to the World Wide Web, fax services, voice mail, electronic mail, paging, and traffic and weather reports. The growing demand for such expanded services is anticipated to push digital phone technology into even higher frequency ranges in the microwave spectrum.

The greatly expanded services offered by digital PCS compared to cellular require the use of broader bandwidth in order to increase to acceptable levels the rate and quantity of information that can be transmitted. Current silicon-based technology cannot always support the necessary microwave technology requirements. Although they historically have been much more difficult and expensive to process and manufacture, GaAs-based MMIC devices are necessary for most microwave transmission applications. The new PCS data-transmission requirements therefore are encouraging widespread commercial development of GaAs MMIC devices for the first time.[3]

Many other new commercial products are increasingly making use of higher-frequency broadband microwave technologies. These include DBS, cable television receivers (cable boxes), "wireless" cable TV systems (28 GHz band), fiber optic communication systems, and many wireless applications. While smaller than the PCS/digital phone market, WLANs make up an extremely important and rapidly expanding commercial sector that is relentlessly driving commercial wireless technology higher into the microwave spectrum.

WLANs permit the interconnecting of personal computers (PCs) without the necessity of additional wiring or cables. WLANs appeal primarily to niche markets where there is a need for networking in a mobile situation where wiring is difficult. Growing market demand for WLAN technology from retail stores, restaurants, medical service providers, warehouses, and other commercial businesses has led to an expected increase in worldwide connections from 1.8 million in 1996 to over 3 million in 1997.

As in the case of PCS/digital phones, the key factor driving LAN technology is the need for higher bandwidth to increase data transmis-

---

[3]See, for example, Hardy (1996) and Arnold (1996).

sion. This factor pushed LAN technology from the 900-MHz range in the early 1990s up to the 2.4-GHz range by the middle of the decade. By the end of the decade the technology is expected to push further up the microwave spectrum to 5.8 GHz (Hostetler, 1996).

AUTOMOTIVE SENSORS

The emergence of automotive ACC and collision warning systems, as well as the projected new generation of LEO and MEO mobile communications satellite systems, are of particular relevance to the future of commercial defense-related RF/microwave technologies. Both these products are promoting the development for the first time of high-technology commercial radar T/R modules that are planned for large-scale, low-cost commercial production. Almost every major automobile manufacturer is investigating a variety of new automotive sensors for ACC, the most interesting of which for our purposes are T/R radar modules that are mounted on front and rear bumpers. Most of the so-called "Big LEOs" are projected to have many communication satellites equipped with up-link/down-link and intra-satellite communication antennas that will be heavily populated with GaAs MMIC-based T/R modules.

Automotive electronics companies and defense microwave component vendors are developing collision warning systems. For example, Delco Electronics Systems, a division of Delphi Automotive Systems, is marketing the Forewarn collision warning systems and developing other integrated ACC systems for passenger cars.[4] Working closely with HE-Microwave, a joint venture between Delphi and Raytheon, a dual-use T/R module facility has been developed that produces products for both the civilian and military markets. The Forewarn systems integrate emerging low-cost microwave and millimeter-wave radar sensors, developed in part from military and aerospace applications, with existing automotive electronics. The Forewarn object detection system has already been widely marketed for school buses. Various technologies have also been tested and integrated on two test passenger vehicles, a Lexus LS 400 and Cadillac Seville. A forward-looking millimeter-wave radar that op-

---

[4]Prior to May 1999, Delphi was majority-owned by General Motors Corporation.

erates in the 77-GHz range has been developed to operate in conjunction with a laser radar. Engineers have experimented with both mechanically scanned antennas and switched-beam radar sensors. The rear detection system is based on sophisticated GaAs MMIC technology and operates in the 24-GHz range, well up in the microwave spectrum. Radar signal-processing techniques are used to discriminate among different categories of targets (Olney et al., 1996).

LOW-EARTH ORBIT MOBILE COMMUNICATIONS SATELLITES

Although the commercial future of many of these systems appears to be in doubt, the new generation of MEO and LEO mobile communications satellite systems are pushing commercial microwave technology up the microwave spectrum closer to technology areas of interest for fire-control radars. If successful, these satellite systems could open up a major new commercial market for active phased-array radar technologies. Unlike automotive collision avoidance systems that use only a few T/R modules per vehicle, next-generation communication satellites will mount large phased-array antennas populated with almost as many T/R modules as on antennas on phased-array fire-control radars.

Most existing communication satellites systems are based on a small number of large geostationary satellites in high-altitude orbits that broadcast a single beam that covers large portions of the world's surface. These satellites are well suited for television transmission and fixed telephone communications. In the mid 1980s, however, engineers began envisioning satellite communication systems that could provide global mobile cellular phone and data transmission capabilities. This concept required much larger numbers of smaller, low-altitude satellites with directional antennas that could transmit many "spot beams" to specific small areas on the earth's surface.[5] This proposed capability required antennas populated with many T/R modules based on GaAs MMIC technology (Kuznik, 1996). In addition, by the early 1990s, market pressures for more data-carrying

---

[5]Geostationary communications satellites are located 22,300 nautical miles above the earth, while LEOs will orbit at altitudes ranging from 420 to 750 nautical miles.

capabilities, combined with growing demand for more broadcasting frequencies, had already been pushing communication satellite transmitters further up the electromagnetic spectrum, from C-band frequencies (4–8 GHz) beyond the X-band into the Ku-band (12.5–18 GHz) (see Figure 3.1). Launched in September 1993, the National Aeronautics and Space Administration's Advanced Communications Technology Satellite (ACTS) provided one of the first successful demonstrations of broadband Ku-band satellite communications (Kuznik, 1996).

Among the most important of the new and projected systems are ICO Global, Globalstar, Iridium, and Teledesic.[6] These are all nongeostationary mobile satellite communications systems intended to provide global real-time voice and data transmission ranging from basic electronic mail to videoconferencing, interactive multimedia and real-time two-way digital data flow. Although several of these systems are now in financial jeopardy—Iridium and ICO Global each filed for bankruptcy protection in August 1999, and Globalstar is now finding it difficult to raise financing—all either do or will deploy advanced active radar antennas. Iridium satellites each have three side panels with multibeam array antennas providing a total of 48 spot beams for mobile users. Feeder uplinks and downlinks operate in the Ka-band. Globalstar satellites each carry planar phased-array antennas for mobile users with 16 spot beams for L-band and S-band coverage. Teledesic satellites will have a multipanel system with many active-element phased-array facets on each panel. Teledesic will operate entirely in the Ka-band. These three systems alone, if fully deployed, will require the design and manufacture of thousands of low-cost, high-performance T/R modules to populate scores of antenna arrays.

Drawing in part on its extensive recent experience in developing military phased-array T/R modules for such programs as the ground-based radar for the U.S. Army's Theater Missile Defense program,

---

[6]Among the major partners on ICO Global are Deutsche Telekom, Ericsson, and Digital Voice Systems, Inc. Globalstar is majority-owned by Loral; other partners include Qualcomm and Alcatel. Raytheon, Siemens, Lockheed Martin, Sprint, Korean Mobile Telecom, and 11 other companies are teamed with Motorola on Iridium. Teledesic was first founded by Microsoft chairman Bill Gates and cable/cellular phone entrepreneur Craig McCaw; Boeing, Motorola, and Matra Marconi Space joined in 1998.

Raytheon has already developed T/R modules for the satellite antennas on the Iridium and Globalstar systems. Raytheon is also deeply involved in the Multifunction Integrated Radio Frequency System (MIRFS) program for development of the next-generation phased-array fire-control radar for the U.S. Air Force/Navy Joint Strike Fighter (JSF) (*The Economist*, 1996).

CONCLUSION

The broader bandwidth requirements of PCS/digital cellular phones, WLANs, DBS, mobile communications satellite systems, and other new consumer broadband technology products have led to an explosion of commercial parts and components development in exotic microwave areas that previously were almost entirely dominated by the military. A dramatic case in point is the widespread commercial development of whole new families of GaAs MMIC and application-specific integrated circuits (ASICs) for the new commercial applications. GaAs MMIC RF power amplifiers and other RF analog devices developed for PCS and other mobile communications may be particularly relevant to future military RF applications, as are some of the manufacturing processes for the T/R modules developed for automotive sensors and mobile communications satellites. Highly specialized GaAs device vendors, such as M/A-Com, Anadigics, and Triquint—companies that for the most part focused heavily on the development of military microwave GaAs devices in the 1980s—have become predominantly commercial foundries for consumer products in the 1990s. They have been joined by commercial giants such as Motorola and Raytheon.

As a result, for the first time, commercial applications are becoming increasingly important in the development of new technologies, especially in lower-cost manufacturing processes for RF/microwave devices. For example, the development of affordable new packaging approaches such as Bump Grid Arrays, Ball Grid Arrays (BGAs), and solder-bumped flip chips for high-frequency MMIC packages are being driven largely by the commercial market for applications in

PCS, WLANs, DBS satellites, and automotive collision-avoidance systems.[7]

Commercial demand for these sophisticated RF/microwave parts and devices is likely to far outstrip military demand. For example, the Teledesic system alone is projected to use many millions of gallium arsenide microchips in its satellites to support RF functions (*Red Herring,* 1996). If ACC and collision avoidance systems become common on passenger cars, automotive T/R module production could rise into the millions.

Many of the parts and components being developed for consumer products may not be directly applicable to or usable in military fire-control radars, nor do they necessarily possess the performance capabilities, ruggedness, and reliability required for the harsh environment in which fighters must operate. Nonetheless, the commercial marketplace is clearly becoming increasingly dominant in broad sectors of RF/microwave technologies and manufacturing processes in a way that could benefit defense applications. Design methodologies, process technologies, and many other areas that have direct relevance to military radar system design and development are likely to be increasingly driven by the commercial market. Commercially developed parts and components will be available for incorporation in military systems, but will such items developed for consumer products possess the high-performance capabilities required for incorporation into weapon systems?

---

[7]An example is the product line being developed by Micro Substrates Corporation.

Chapter Four

COMMERCIAL INSERTION AND THE QUESTION OF WEAPON SYSTEM PERFORMANCE

INTRODUCTION

As noted in earlier chapters, many CMI advocates argue that market-driven commercial R&D has surpassed military R&D in diverse technology areas. The implication is that the broader use of commercially developed technologies in the military sector would result in more-capable weapon systems. Yet paradoxically, perhaps the single most important and deeply felt concern expressed by CMI skeptics is that the military use of commercial parts, components, designs, and technologies will result in less-capable and less-reliable weapon systems. These concerns arise from a belief that weapon systems often must be able to operate in far more stressful and demanding environments than commercial products. This is a crucial issue, because CMI must be viewed as an unacceptable strategy if it results in less-capable or less-reliable weapon systems than are needed by America's armed forces.

This chapter attempts to shed additional light on the debate over dual-use performance by examining evidence from our case studies of various aspects of CMI applied to RF/microwave military technology programs to answer our second question:

- Is the commercial market driving technology at a rate and in a direction that meets national security requirements? In other words, can CMI provide the necessary and desired military equipment performance capabilities?

We first examine the question of "insertion" of commercially developed parts and components into radars and other RF/microwave military systems. We then briefly examine the use of commercially derived technologies and designs in similar systems.[1]

INSERTION OF COMMERCIAL PARTS AND COMPONENTS

Parts insertion refers to the use of standard "off-the-shelf" (OTS) commercial electrical components and parts such as integrated circuits in weapon systems. Prior to June 1994, almost all parts used in military systems were required to adhere to official Mil-Specs. Mil-Specs were originally developed in the 1960s to ensure that parts used in military systems would possess the necessary robustness and capabilities to operate in harsh military environments, and to ensure reliability, quality, configuration control, and logistics support.[2] By the early 1990s, approximately 40,000 Mil-Specs provided uniform technical and management standards for the design and development of weapon systems and other military applications. Technical Mil-Specs can be very precise and detailed, often specifying materials, processes, design standards, and so forth down to the lowest parts level.

As noted in Chapter Two, acquisition reformers argue that Mil-Specs are a major barrier to CMI, because commercial parts and components are usually designed and developed in accordance with different technical standards. As a consequence, Mil-Specs often require the use of unique parts specially developed for military applications. Such parts are produced in small quantities and must be subjected to extensive test and screening procedures, and thus are very expensive. Yet acquisition reformers contend that Mil-Specs are often outdated, sometimes mandate the use of unnecessary capabilities and technologies, and may lead to the use of less-capable but more-expensive technology than is available in the commercial sector.

---

[1]DoD has defined "insertion of commercial capabilities into military systems" as the "Third Pillar" of its three-pillar dual-use CMI strategy. According to DoD, insertion entails the use of "best commercial materials, products, components, processes, practices, and technologies into military systems whenever possible." See OUSD/A&T (February 1995).

[2]This discussion of Mil-Specs is drawn largely from OUSD/A&T (October 1996), Appendix A.

Nonetheless, serious concerns remain in the services and in the defense industry regarding the advisability of eliminating Mil-Specs and routinely using OTS commercial parts in military systems.[3] These concerns arise from the view that the military environment is far harsher and more demanding than the relatively benign environments in which most commercial parts are designed to operate. The most serious concerns focus on the capabilities of commercial parts to withstand the harsh extremes of temperature, vibration, altitude, g-forces, and humidity required by Mil-Specs. Observers also have raised questions about cycle and long-term shelf life.

As a result of these concerns, DoD and the services have concluded that initial Mil-Spec reform implementation was "frequently overzealous," and that greater caution and care must be taken in eliminating Mil-Specs. Studies have been undertaken to determine more precisely where the continued requirement for Mil-Spec parts might be warranted.[4] These studies have contributed to a growing body of evidence that, if implemented with care, insertion of a wide spectrum of selected commercial parts in RF/microwave military systems and other military electronics can be a viable strategy.

In 1996, the Electronic Industries Association (EIA) surveyed eight major military electronics contractors to determine the use of commercial parts in military systems.[5] Seven of the eight participating firms had experience with applying commercial parts to military systems, including radars, missiles, and communications systems.[6] The principal conclusion of the study was that commercial parts can be incorporated in military systems without significantly degrading system performance.

---

[3]For a thorough and thoughtful review of concerns in the defense electronics industry, see EIA (January 1997).

[4]See, for example, OUSD/A&T (October 1996).

[5]The eight participants were Allied Signal, David Sarnoff Research Center, GEC-Marconi Electronic Systems, GTE, Lockheed Martin, Northrop Grumman, Rockwell International, and Texas Instruments.

[6]Applications included MODAR, APS-134(LW), APS-137(V), F-16, Joint Tactical Information Distribution System and Multifunctional Information Distribution System, Longbow missile, Tomahawk, Javelin Command Launch Unit, Wind Corrected Munitions Dispenser, and the Tri-Band Tactical Terminal.

However, a major caveat accompanying this conclusion was that the data supporting it are still "limited." The single greatest concern of the respondents was the difficulties they were experiencing in finding adequate characterization data on commercial parts so that they could be used with confidence as substitutes for Mil-Spec parts. There is a lack of information about the long-term reliability of commercial parts when used in stressful military operating environments. In other words, a commercial off-the-shelf (COTS) or custom part might substitute for a Mil-Spec part perfectly well in performance characteristics, but might not—it was feared—possess the robustness to provide the required reliability in harsh military environments. The key technical issues were:

- Temperature operating ranges

- Tolerance to moisture

- Tolerance to vibration

- Tolerance to high-altitude environments

- Tolerance to high g-forces

- Cycle life

- "Footprint" incompatibilities.[7]

A wide range of grades of COTS and custom parts that are not Mil-Spec are available on the market.[8] These grades represent a spectrum of parts with different temperature, moisture, and vibration ranges that are available in the commercial world. The most com-

_____

[7]The footprint of an electrical part refers to its size, number of pins, pin arrangement, and so forth.

[8]Industry parts specialists argue that, strictly defined, all parts that are available from standard catalogs, including Mil-Spec parts, are COTS parts. Standard Mil-Spec catalog parts are sometimes called Government Off-the-Shelf (GOTS). Custom parts are COTS or GOTS parts that require additional screening, testing, or selection beyond the catalog definition. Custom-designed parts, whether used in military or commercial applications, are not considered COTS parts. Many military radars use nonstandard or custom-designed parts that must receive government approval. These parts technically are not Mil-Spec but rather government-approved nonstandard parts. Some industry experts argue that the use of Mil-Spec GOTS parts is often cheaper than using non–Mil-Spec COTS parts, because of the high costs of screening the latter parts for insertion into military systems. See Martin (1995).

mon grades of commercial electronics parts are called "consumer grade," "industrial," or "automotive grade," as shown in Table 4.1. Thus, industrial-grade parts, which are used in automobiles and trucks, have much wider recommended temperature operating ranges than do consumer-grade parts, which are meant for consumer electronics such as televisions and VCRs. Also, screening (testing beyond catalog definition of recommended performance environment ranges) is often available for industrial parts, although usually not for consumer parts. Indeed, most major car manufacturers have close relationships with their electronics parts vendors and set various rigorous standards for operating environments, which may require screening. In a variety of demanding commercial applications, such as engine-control integrated circuits for heavy construction equipment, for example, parts may be screened to operate in temperature, vibration, heat, and moisture environments that exceed those typical of Mil-Spec parts. With regard to footprint, commercial parts are often incompatible with Mil-Spec parts because the rapidly advancing commercial market has pushed the parts technology beyond the Mil-Spec world.[9] If military electronics modules are designed for commercial parts insertion from inception, footprint is usually not a problem.[10]

Table 4.1

Examples of Differences Between Various Grades of Parts

Characteristic	Mil-Spec Grade	Consumer Grade	Industrial/ Automotive Grade
Temperature range	−55°C to +125°C	0°C to +70°C	−40°C/−25°C to +85°C
Packaging/ encapsulation	Ceramic or metal	Plastic	Plastic or hermetic
Screening	Yes	Usually none	Usually none, but available
Footprint	Mil-Spec baseline	Usually incompatible	Usually incompatible

[9]Some differences also arise from encapsulation in plastics vs. ceramics or metals.

[10]Severe space and weight constraints in densely packed fighter aircraft sometimes require specialized parts, packaging, and cooling.

Our case study evidence suggests that the real-world experiences in system performance of contractors committed to commercial parts insertion have been mostly positive. A key characteristic of successful efforts appears to be contractor configuration control; that is, granting the contractor the freedom (and responsibility) to select the most cost-effective and appropriate grade for every part in an avionics module, assembly, or system.

Mil-Specs tend to be extremely conservative and sometimes grossly overspecify performance-range requirements for parts. Engineers who are actually designing and developing a specific module often have a far better understanding of what performance requirements are necessary for each part and component in that module. Mil-Specs can limit or constrain cost-effective solutions by overspecifying requirements or mandating the use of outdated or inappropriate technologies. And there is a direct correlation between parts grades and cost, as illustrated in Figure 4.1. Sometimes an industrial-grade or even a consumer-grade part may be perfectly adequate for a given

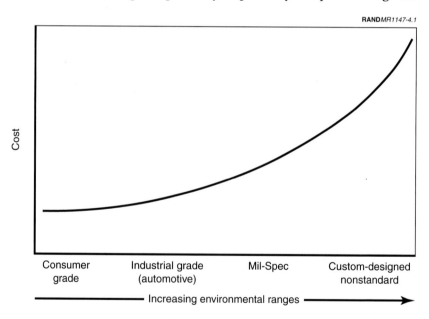

Figure 4.1—Greater Temperature, Moisture, and Vibration Ranges
Increase Parts' Costs

application and may adequately support the overall performance capabilities requirements of the module. At other times, no commercially available part exists—or none that has the necessary environmental performance range characteristics—for a specific military application, particularly in the RF/microwave world. In such cases, Mil-Spec or other types of nonstandard or custom parts should of course be used.

The limited experience of contractors suggests that a carefully selected mix of Mil-Spec, industrial-grade, and consumer-grade parts can be used with little or no degradation in performance and high payoffs in lower costs, provided the contractor is granted configuration control and change authority. An interesting recent experiment in this area is the Manufacturing Technology Industrial Base Pilot Program (Mantech IBP) for producing military products on commercial production lines.[11] Sponsored by the U.S. Air Force Manufacturing Technology Directorate at Wright Laboratory, the Mantech IBP program demonstrates the design and manufacture of complex military avionics components in accordance with best commercial practices. TRW Avionics Systems Division has been contracted to produce two electronics modules from the CNI for the F-22 fighter and the U.S. Army's Comanche RAH-66 reconnaissance/attack helicopter. These modules will be manufactured by TRW's Automotive Engineering Group at a standard, fully automated commercial automotive electronics plant in Marshall, Illinois.

A key objective of the program is to maximize compatibility with normal automotive production-line processes and minimize disruption to ongoing commercial programs. To achieve this objective, the military modules had to exhibit maximum parts and design commonality with the commercial automotive items produced at the factory such as electronic engine controls and air bag sensors. At the same time, the modules had to possess equivalent performance capabilities and cost less than modules manufactured on military lines, and had to be fully compatible with the other modules in the F-22 and RAH-66 CNI. These objectives had to be achieved without electrical redesign of the modules. To achieve compatibility with auto-

---

[11]For an overview of this program, see U.S. GAO (June 1996).

motive electronic manufacturing processes, Mil-Spec parts had to be replaced with commercial automotive-grade parts.[12]

Engineers selected two typical digital CNI modules—out of a total of 38 modules that make up the complete CNI system—for the pilot program: The RF/front-end controller and the pulse narrowband processor. These modules were reconfigured so that they contained about 90 percent commercial parts and components. The redesigned modules used the advanced plastic packaging approaches and recently developed ball grid-array technology for ASICs and multichip modules that are used in cutting-edge commercial applications, but that are virtually non-existent in military electronics.

Initial testing indicates that the performance, reliability, and cycle life of these modules equals or exceeds similar modules built with full Mil-Spec parts and components and manufactured on Mil-Spec military production lines. As of mid 1996, initial testing showed that the functionality and durability life of the two IBP test modules equaled the baseline performance established for the Mil-Spec modules. Plastic encapsulated microcircuits passed all tests. Test results also showed that large ball grid arrays used in the modules could endure stressful thermal cycling. Furthermore the IBP test modules weighed about 15 percent less than the baseline target weight for the Mil-Spec modules.[13]

We identified two other interesting cases of commercial parts insertion: The AIL Systems family of modular radars derived in part from the AN/TPS-74 Modular Radar (MODAR) developed for the U.S. Army in the late 1980s, and the Northrop Grumman ESSD/GEC-Marconi Systems Programmable Digital Radio (PDR) for CNI applications.

---

[12]As noted earlier, the different footprints of Mil-Spec parts often make them incompatible with standard commercial automated manufacturing equipment.

[13]TRW engineers argue that plastic encapsulated microcircuits are suitable for most military applications. See Myers and Bartlett (1996). An alternative view can be found in Donlin (February 1995). Donlin raises concerns over the lack of data on the reliability of plastic encapsulated microcircuits when used in systems such as missiles that are stored in a dormant state for years in harsh environments or subjected to high humidity levels.

AIL is developing and producing a limited production run of radar common modules, including antennas, receiver/exciters, power amplifiers, pedestals, and signal processors. The company is targeting a wide range of potential U.S. and foreign military and commercial customers, so the modularity of the design is critical. The modules can be mixed and matched to produce radar systems appropriate for a wide range of military and civilian ground and airborne applications.[14] AIL originally intended to procure all full-Mil-Spec-grade parts. However, cost and schedule considerations led AIL to consider industrial/automotive- and consumer-grade parts. For many components, AIL engineers concluded that the use of non–Mil-Spec parts would provide equal or better performance capabilities at a lower cost with a shorter development time.[15] The resulting radar modules are a mix of Mil-Spec, industrial-grade, and consumer-grade parts.

Because of continuing uncertainties regarding the reliability of non–Mil-Spec parts in harsh military environments, AIL recognized the importance of implementing a thorough test program of its modules. Engineers are paying special attention to rigorous temperature and vibration testing of the modules. Typical are the results obtained with the signal processor, which contains all three grades of parts, and the pedestal, which contains all industrial-grade parts. No vibration problems were encountered with either module during testing. The same was true when the modules were tested at temperatures well above the recommended ranges for industrial- and commercial-grade parts. However, serious performance degradation problems were encountered at temperatures below –30°C. For effective operation in environments below –30°C, the modules will have to be protected or different parts will have to be used.

Northrop Grumman ESSD and GEC-Marconi are developing a flexible PDR based on a similar modular concept, for use in a variety of applications, including CNI upgrades for export versions of the Boeing/McDonnell F-15. The PDR will eventually incorporate a va-

---

[14]Existing versions include a Ku-band ground-based moving target indicator battlefield surveillance radar (AN/TPS-74) and Ku-band Remotely Piloted Vehicle/Unmanned Aerial Vehicle (UAV) modular radar.

[15]In general industrial- and consumer-grade parts can be obtained much more quickly from suppliers than can Mil-Spec parts. See Chapter Five.

riety of microwave functions, including Ku/Ka-band satellite communications and microwave landing systems.[16] Similar to the AIL MODAR, the PDR uses non–Mil-Spec parts whenever possible to keep costs down and shorten the R&D cycle. When Mil-Spec parts were deemed necessary, GOTS components were selected. For example, the system uses an autonomous target-recognition chassis and card set from a proven militarized radar product, combined with industrial- and consumer-grade parts such as COTS television filters and field-programmable gate arrays. Overall the developers applied normal commercial ISO 9000 standards rather than Mil-Spec standards. The PDR prototypes have been successfully tested in simulated operational environments such as at the 1995 Joint Warrior Interoperability Demonstration in San Diego.

Other experiments are under way by a wide variety of contractors working with all three services to examine the insertion of non–Mil-Spec parts into military systems and subsystems. Our examination of several case studies in the demanding RF/microwave technology area provides additional encouraging—though still limited and qualified—evidence that a carefully framed CMI strategy on the parts level can lead to systems exhibiting equal or better performance capabilities compared with systems developed entirely with Mil-Spec parts.

Several caveats are necessary, however. There still are not analogous parts and components available on the non–Mil-Spec commercial market necessary for certain microwave applications used for advanced military radars. Yet, as discussed earlier, civilian applications are now increasingly driving RF/microwave and MMW technologies, leading to the commercial development of more and more parts and technologies relevant to military microwave applications. Defense contractors interested in CMI have adopted a strategy of inserting a mix of custom Mil-Spec, industrial-grade, and consumer-grade parts into RF/microwave subsystems. The specific mix is determined by tradeoff analyses of a variety of factors such as overall system performance, cost, and schedule requirements.

---

[16]The baseline prototype design is 1.5 MHz to 2.8 GHz.

Even with the current approach of using a mix of different grades of parts, concerns remain widespread in industry regarding the long-term reliability and durability of such hybrid systems. The test evidence is mostly encouraging, but it remains limited. Hybrid systems will have to continue to be tested extensively for their long-term resistance to extremes of temperature, humidity, vibration, and other environmental factors before they can be used with complete confidence in stressful military environments.

INSERTION OF COMMERCIALLY DERIVED DESIGN APPROACHES, TECHNOLOGIES, AND PROCESSES

The simplest form of CMI entails insertion of non-Mil-Spec COTS parts and components into military systems. A more comprehensive CMI strategy would involve "spinning-on" or more effectively exploiting commercially derived design approaches, technologies, and processes for military applications. Such a strategy would not necessarily require the use of COTS parts, although such an approach would be encouraged. Rather, developers of military systems would attempt to take greater advantage of relevant design approaches and technologies available or under development in the commercial marketplace. CMI advocates claim that such an approach will improve system performance through the incorporation of more-advanced commercial technologies and processes into weapon systems, reduce R&D costs by piggybacking on commercial R&D expenditures, and help maintain a dual-use industrial base at lower cost to the government. Several of the case studies we examined provided encouraging, though limited, evidence supporting this claim of the CMI advocates.

DARPA has been particularly active in sponsoring projects that promote dual-use technology development in microwave technologies and other areas. One project of direct relevance to our study is the Technology Reinvestment Project (TRP) on RF/Microwave/MMW technologies, supervised by the U.S. Air Force Wright Laboratory.[17]

---

[17]The official program title is "Development and Application of Advanced Dual-Use Microwave Technologies for Wireless Communications and Sensors for IVHS Vehicles," but the scope of the effort has been broadened considerably since this title was formulated. The participants are Northrop Grumman Electronic Systems, M/A-

A goal of this program is to help defense contractors leverage their skills and capabilities in military microwave technologies to enter commercial markets in related areas, to promote the "spin-back" of more advanced commercial technologies to defense applications.

Northrop Grumman Electronic Systems in Rolling Meadows, Illinois, is a major participant in the DARPA/TRP programs. This company is the developer and manufacturer of the ALQ-135 electronic warfare system deployed on the Boeing/McDonnell F-15E, and is involved in a variety of other EW programs that draw heavily on microwave and MMW technologies. Under the auspices of the DARPA TRP program launched in 1994, Northrop began to develop a variety of commercial "spin-off" applications of its microwave technologies in automotive radar sensors and wireless communications. After two years, Northrop had developed a variety of systems and components, including a 24-GHz Wireless Link GaAs MMIC transceiver module, 900- and 1800-MGz MMIC-based wireless modems, a 24-GHz automotive radar sensor, radio-frequency identification systems, and a wideband 2–6-GHz Microwave Power Module (MPM), and had begun development of an 18–40-GHz MPM (Northrop Grumman, July 1996).[18]

Development of these commercial applications had an immediate and dramatic effect on the contractor's military product development plans. The development of advanced solid-state wideband MPMs for commercial use is of particular interest for military microwave applications. The most technologically advanced EW systems deployed on U.S. Air Force fighters still use large, extremely expensive, low-reliability traveling wave tube technology. For microwave transmitter applications, the commercial world has moved toward much more reliable, cheaper, lighter-weight MPMs for high-power amplifiers.

TRP program experience developing commercial technologies has led to new proposals for incorporation of advanced MPM technology into EW and other military system applications. As mentioned ear-

---

COM, Wright Laboratory, University of Illinois, Northwest University, and several smaller private companies.

[18]Related technology developments came out of the DARPA-sponsored Microwave and Millimeter-Wave Integrated Circuit (MIMIC) and Microwave and Analog Front-End Technology (MAFET) programs .

lier, ball grid-array integrated circuit packaging technology has been driven by commercial developments, but is virtually non-existent in Mil-Spec electronics applications. The TRP program led participants to develop and employ ball grid-array technology for commercial wireless applications, then make it available for military applications. The same is true of many other advanced commercial microwave technologies, such as direct-sequence spread-spectrum devices with ASIC/MMIC, various forms of plastic packaging, and many process technologies.

CONCLUSION

Our findings on the performance consequences of commercial technology insertion into military avionics, based on analysis of our case studies, can be summed up as follows:

- Limited evidence suggests that commercial-grade parts and components can be successfully inserted into RF/microwave military avionics systems without degrading system performance. However, many Mil-Spec, specially screened commercial parts, or custom-designed parts and components are still likely to be necessary. Furthermore, legitimate concerns remain about the long-term reliability and durability of commercial-grade parts. These concerns can be addressed through further testing and experimentation.

- Limited evidence suggests that commercially derived designs, technologies, and processes can be successfully applied to military RF/microwave systems with the potential of increasing performance. Many of the design approaches and technologies in the commercial sector appear to be far more advanced than what is currently available in the military sector.

- Granting full configuration control and change authority to the contractor appears to promote the successful insertion of commercial parts and technologies into military RF/microwave systems.

Chapter Five

DUAL-USE TECHNOLOGIES: IMPLICATIONS FOR COST, SCHEDULE, AND CONTRACTOR CONFIGURATION CONTROL

INTRODUCTION

Possibly the single most important claim of CMI advocates is that closer integration of the military and commercial industrial bases will lead to significantly lower-cost weapon systems that will be developed more quickly. As has been pointed out in the previous chapter, limited evidence suggests that equal or better performance is obtainable through the use of commercially derived parts and technologies in military RF/microwave systems. At the same time, our examination of case studies also indicates that risks are incurred in moving toward a full-blown CMI strategy, particularly with respect to durability and reliability. These risks are at least partially offset by the promise of much reduced weapon system costs. In this chapter, we examine some of our case study evidence to determine if significant cost-savings and schedule benefits are really likely as a result of dual-use products and technologies. Once again, we divide our analysis into two parts: Parts insertion and technology insertion. Finally, we discuss more fully the question of the importance of contractor configuration control throughout the life-cycle of a system for the successful implementation of CMI.

INSERTION OF COMMERCIAL PARTS AND COMPONENTS

AIL has generated considerable data during the development of its Modular Radar prototypes on the potential cost and schedule benefits of using commercial-grade parts and components.[1] One reason AIL engineers dropped their original plan to use all Mil-Spec parts was because the far-shorter delivery times for commercial-grade parts shortened the schedule and thus the cost for development of the prototype radars. AIL discovered that deliveries of Mil-Spec parts often took six to nine months. Mil-Spec parts that were available in catalogs (GOTS parts) were often not kept in stock. Producers did not keep many Mil-Spec microwave parts in stock and did not produce for inventory. Rather, factories routinely waited for sufficient orders to come in to justify startup of a new production run. Mil-Spec suppliers also tended to arbitrarily discontinue a part at any time with little advance notice.

AIL found that delivery schedules for industrial- and consumer-grade parts were much shorter. Delivery of industrial-grade parts took four months or less. If consumer-grade parts were kept in stock, delivery schedules were even shorter. However, delivery schedules could be as high as six months if the parts were not in stock. On the negative side, commercial vendors usually required minimum buys and would either not sell at all or would charge much higher prices for smaller quantities.

Figure 5.1 gives two examples of the differences in schedule and cost for Mil-Spec and commercial-grade parts. The left side of the figure compares prices for a Mil-Spec and an industrial-grade Pulse Compression Network, a custom-designed RF part. Two part versions are shown, the Dash-1 and Dash-2. The industrial-grade and Mil-Spec versions of the part are identical in performance but not in recommended temperature range, resistance to humidity and vibration, and so forth. The industrial-grade parts are about 40 percent cheaper than the Mil-Spec parts. Furthermore, the industrial-grade parts take one-third less time for delivery. Figure 5.1 also compares the price of a custom-designed Mil-Spec power-supply component with a consumer-grade component with the exact same design and

---

[1]See Chapter Four for a more detailed discussion of this program.

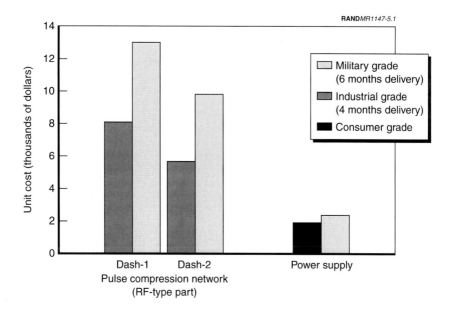

Figure 5.1—Schedule and Cost for Mil-Spec and Commercial-Grade Parts (Pulse Compression Network)

performance characteristics. The consumer-grade component costs about 20 percent less.

For many years, numerous Mil-Spec electronics parts have been manufactured on dual-use commercial lines and are in fact identical to commercial parts. However, these parts may have enormous price differences because of the extra screening and testing required of each Mil-Spec part. In the commercial world, manufacturing processes or specific vendors—but not each and every part they produce—are often qualified by the system integrator.[2] In contrast,

_____

[2]Sometimes a commercial parts vendor is required by a buyer to qualify each part through a specific process. In the commercial world, the integrator usually works with the parts vendor prior to mass production and takes part in the testing of prototype parts. Further, many parts purchasing agreements in the commercial arena include provision for financial rewards and punishments with respect to timeliness of delivery, parts quality, and so forth.

each Mil-Spec part is subjected to rigorous testing, which greatly increases its cost.

Figure 5.2 shows the basic 10-part lot cost for two parts investigated by AIL for their Modular Radar program, plus the cost of screening. Engineers looked at two RF mixers: one Mil-Spec and one consumer grade. The basic 10-part lot cost for both is $410. However, for the Mil-Spec version, the vendor adds a lot charge plus $15,000 for screening the parts. Whereas the commercial RF mixer was in stock and immediately available, the Mil-Spec version required a minimum of four months for delivery.

AIL also investigated using two Mil-Spec digital integrated circuits (ICs) (750-1, 751-1) in their modular radars (see Figure 5.2). The vendor had discontinued manufacture of the Mil-Spec parts, but the nearly identical consumer-grade ICs were available for $10–$20 each. To deliver the Mil-Spec part, the vendor asked for $121 for the die per

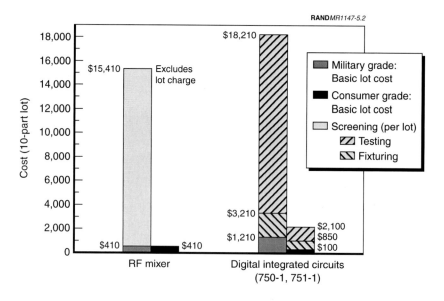

Figure 5.2—Large Cost Premiums Are Paid for Mil-Spec Parts Screening

IC, plus $2000 for fixturing[3] and $17,000 for hermetically repackaging and testing the IC. AIL decided to buy the consumer-grade parts, which are encapsulated in plastic, and conduct its own limited temperature tests. This testing cost $750 for fixturing and $1250 for lot testing. By adopting this approach, AIL was able to purchase a small lot of 10 parts for less than one eighth the cost of a 10-part Mil-Spec lot. The consumer-grade ICs were then inserted into the prototype radar modules, which are themselves being further tested for durability in harsh environments.

The U.S. Air Force/Wright Laboratory IBP program discussed in Chapter Four for developing and manufacturing lower-cost modules for fighter and helicopter CNI systems also demonstrates the cost savings that can occur through the insertion of commercial-grade parts and the manufacture of military avionics components on higher-volume automated dual-use production lines. After the maximum insertion of commercial parts, the two CNI modules are estimated to cost only about 60 percent of the original F-22/RAH-66 baseline cost projection. It is likely that the projected cost would be even lower if the modules had been designed from inception for the insertion of non–Mil-Spec parts. It will be recalled that the program did not permit basic electrical redesign of the modules. In part because of this restriction, 10 percent of the parts remained Mil-Spec. These 10 percent, however, accounted for 50 percent of the module cost.[4]

In sum, limited evidence from case studies indicates that use of commercial parts, when feasible, results in dramatic cost savings. Most commercial parts would have to be screened and possibly

---

[3]Fixturing includes the costs of setting up rigs and making other special hardware required for custom testing.

[4]A breakout of costs by the General Accounting Office (GAO) suggests that the bulk of the estimated savings result from the insertion of commercial parts and the manufacture of the modules on a standard (low-volume) commercial production line. According to the GAO, about 40 percent of the cost savings arise from reduced labor costs resulting from automated commercial manufacturing facilities. Twenty percent of the savings is attributed to less-expensive materials, and about one quarter to the elimination of military specifications and standards that called for special testing, screening, and other material compliance measures. The remaining 20 percent of savings is attributed to reduced administrative burden attributable to the relaxation of standard defense acquisition oversight measures. See GAO (June 1996, pp. 4–5).

ruggedized or repackaged prior to use in military systems, but it appears that even with these caveats commercial parts may often be less expensive than Mil-Spec parts.

INSERTION OF COMMERCIALLY DERIVED DESIGN APPROACHES, TECHNOLOGIES, AND PROCESSES

Part of the cost savings projected for CNI modules from the U.S. Air Force/Wright Laboratory IBP pilot program arise from the application of commercial manufacturing processes to the production of the modules. Several of our case studies indicate that insertion of commercial technologies and processes in other areas could lead to significant cost savings in military-specific RF/microwave avionics.

For several years Wright Laboratory has sponsored various radar technology demonstration and pilot programs that encourage the incorporation of commercial technologies and techniques directly into military aircraft radars. Two such programs are the Advanced Low Cost Aperture Radar Program (ALCAR) and the Radar System Aperture Technology Program (RSAT).

One purpose of these two programs is to promote the development of much lower-cost technologies for phased-array fire-control radars. The participating contractors have examined a wide variety of strategies to reduce costs while maintaining system performance. These strategies include assessment of different technical approaches based on commercially developed technologies. A complementary program, the Multifunction Integrated Radio Frequency System (MIRFS) program, sponsored by the Joint Strike Fighter Program Office (PO), is looking at similar questions for development of the next-generation U.S. Air Force fire-control radar.[5]

As mentioned in Chapter One, a key cost driver in new-generation fire-control radars is the high cost of T/R modules for electronically scanned antenna arrays. Pilot programs are examining different techniques and design approaches to solving this problem. On the RSAT program, Raytheon has developed a completely new low-cost antenna architecture and technology that was originally developed

---

[5]Both Raytheon and Northrop Grumman participate in all three of these programs.

for commercial applications. Raytheon calls its new approach the Continuous Transverse Stub (CTS) Electronically Scanned Array concept. Commercial uses under consideration for CTS technology include antennas for DBS TV. For airborne military radar applications, the CTS concept replaces a planar-array aluminum antenna with a much less expensive array manufactured from common extruded plastic. No expensive machining or milling is required. Using other innovative mounting and phase-shifting techniques, the CTS antenna can be combined with a relatively small number of T/R modules to produce a low-cost active array.

Another approach examined by Northrop Grumman for a low-cost ESA was to exploit rapidly evolving commercial technology developments in MPMs to develop a lower-cost MPM-based transmitter as an alternative to expensive solid-state transmitters or low-reliability traditional travelling wave tubes. These MPMs are used as building blocks for a modular design architecture for the antenna that is projected to result in a much lower-cost array. The basic technological approach adopted to achieve this effect has been widely applied throughout commercial industry. As shown in Figure 5.3, the resulting redesign of the antenna results in significantly fewer parts and components at much less cost.

Both of the new technology approaches discussed, however, lead to somewhat lower-performance radar antennas when compared to more-traditional arrays populated with large numbers of T/R modules. Several radar developers have been examining approaches to reducing the manufacturing costs of traditional T/R modules and related high-cost microwave components and assemblies. Many of these approaches include incorporation of technology or design approaches first developed in the commercial sector. As a result of such efforts, the costs of military T/R modules have declined dramatically since the beginning of the 1990s. Figure 5.4 shows a generic curve that describes the typical cost reductions that have been achieved by the leading producers of military T/R modules. These cost reductions are in the range of an order of magnitude.

Many factors are responsible for these enormous decreases in average unit cost, including increased automation in assembly, reduced MMIC costs, new technology insertion, and greater use of commer-

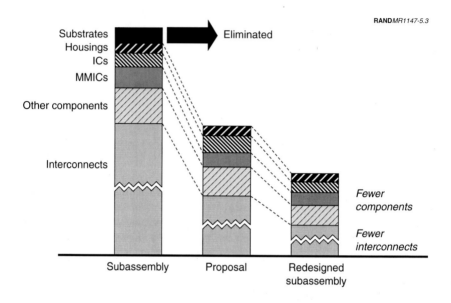

**Figure 5.3—Nontraditional Technology Approaches and Commercial
Spin-On Can Reduce Antenna Complexity and Costs**

cial parts and technologies. For example, costs have been reduced
through insertion of commercially developed parts into T/R modules
such as low-noise amplifiers drawn from direct broadcast television
systems. Closer adherence to commercial design rule practices have
contributed to cost reductions. Insertion of new technologies developed for dual-use applications, such as aluminum nitrate substrates
and silicon germanium wafer processing, have helped to bring costs
down.

Two other examples are of particular interest in regard to T/R modules. In one case, a contractor's military division worked closely with
an automotive commercial electronics division to improve manufac-
turability and yield. As a direct result of the interaction with the
high-volume commercial electronics division, the military division
redesigned its T/R module to reduce the number of wire bonds, decrease the number of chips on a single substrate, and separated the
GaAs and Si chips onto separate substrates.

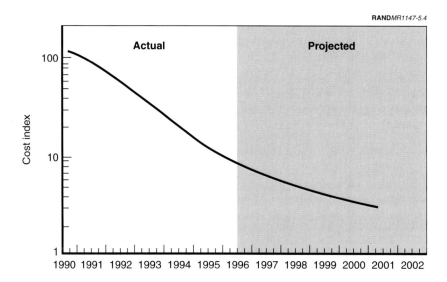

Figure 5.4—Typical T/R Module Cost History and Projection

In another instance, a defense division, after interacting with a commercial electronics division, decided to adopt "flip chip" technology for military high-power microwave applications. This technology is common in the consumer electronics world in various straight digital logic applications, but had never been used before in military microwave applications. The advantage of attempting to apply commercial flip chip technology to microwave applications was an increase in the thickness of the chip and elimination of wire bonding to permit manufacture on high-speed automated equipment and to increase yield and reliability.

The cases related here are just a few examples among many that we encountered. As a result of our analysis of the case study evidence, we conclude that:

- The systematic insertion of commercial parts, technologies, and manufacturing processes, combined with dual-use automated manufacturing, is likely to reduce the costs of typical military avionics modules by roughly 20 to 50 percent, and to shorten R&D schedules. Cost-saving potentials appear to be greater in digital avionics than in high-end RF/microwave applications, but this may change as commercial microwave applications become more widespread.

THE ROLE OF CRADLE-TO-GRAVE CONTRACTOR CONFIGURATION CONTROL

Our examination of the case studies has led us to conclude that contractor configuration control "from cradle to grave" may be necessary to realize the full potential benefits of CMI, for three reasons:

- During R&D, the contractor may need maximum flexibility to select the optimal mix of grades and types of commercial (and Mil-Spec) parts and technologies, as well as design approaches, to achieve the desired performance at minimum cost.

- Limited evidence suggests that a substantial percentage of the projected cost savings from insertion of commercial parts and technologies flows from a strategy of continuous technology insertion over the lifetime of the system. Contractors granted long-term support contracts at a prenegotiated fixed price and configuration control may have strong financial incentives to reduce costs through continuous insertion of more-capable, lower-cost technologies available from the commercial sector.

- The problem of diminishing manufacturing sources for Mil-Spec parts, combined with greater use of commercial-grade parts, means that military electronics will be increasingly affected by the short life-cycles and rapid turnover of commercial electronics technology. Long-term support of military equipment may require a policy of continuous insertion, which may be most efficiently handled through granting configuration control and change authority to the contractor during and after system R&D.

The first point has been made in Chapter Four and elsewhere. The lack of complete and easily accessible characterization data for commercial-grade parts, the need to make thousands of cost-benefit-schedule tradeoff decisions down to the lowest parts level during the design process for each module, and the need to experiment and test mixes of different parts grades in new combinations throughout the design and development process, all suggest that the design engineers actually developing the module need to be granted maximum configuration control and change authority during R&D for the potential benefits of CMI to be realized. Performance and reliability requirements, and form, fit, and function (FFF) parameters need to be provided to the contractor. The efficient incorporation of commercial parts and technologies at minimum cost seems unlikely unless the contractor is granted significant configuration control and change authority at the module level, along with increased responsibility for outcomes.

The second point is illustrated by a major bid submitted in 1995 by a leading avionics contractor for modernizing key RF/microwave areas of existing U.S. Air Force fighter avionics suites with a new series of avionics modules. The contractors' total life-cycle cost estimate was about one-third the baseline estimate generated by the U.S. Air Force, as shown in Figure 5.5.[6] The contractors' lower estimates were based on the following assumptions:

- Maximum insertion of commercial designs, parts, and technology into the new avionics modules.

- Manufacture of the modules on a commercial electronics production line.

- Contractor configuration control and change authority for design and development of the modules.

- Contractor logistics support and depot maintenance for the life of the system.

The costs of installing the new avionics modules remained about the same for both the baseline Air Force estimate and the lower CMI estimate. However, as shown in Figure 5.6, the contractors' estimated cost of the modules themselves was less than 50 percent of the baseline estimate. The CMI estimate included O&S and EMD costs of about 15 percent less and 35 percent less, respectively, than the baseline estimates. The Air Force assessment of the contractors' bid concluded that it was largely reliable and credible.

------

[6]The cost estimates included R&D, production and installation, and operations and support (O&S).

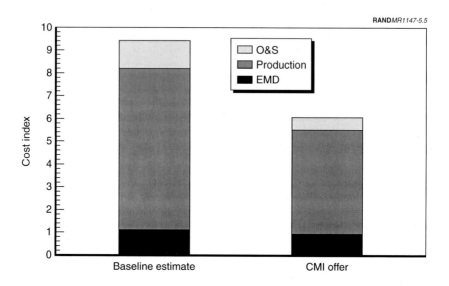

**Figure 5.5—Baseline vs. CMI Life-Cycle Cost Estimates for
Fighter Avionics Upgrade**

The avionics contractor maintained that its lower-cost CMI bid was highly dependent on: (1) contractor support of the system throughout its life-cycle, and (2) full contractor configuration control and change authority. The much lower CMI bid was based in part on projections of trends in costs and capabilities in commercial integrated circuits and other electronics technology over the next decade. The contractor believed that winning full configuration control and change authority would permit continuous insertion of lower-cost, higher-capability parts into the modules. In this way, the cost per module could be reduced by 50 to 70 percent over 10 years. In addition, the number of modules necessary to perform the same function could be reduced drastically, thus further reducing costs or permitting greater capability at the same cost.

The motivation for the contractor would be a prenegotiated fixed-price support contract for a 10-year period. The contractor believed that it might only break even at best on many of the initial modules, but with new technologies constantly arising in the commercial sector, the contractor would have the motivation and the authority to

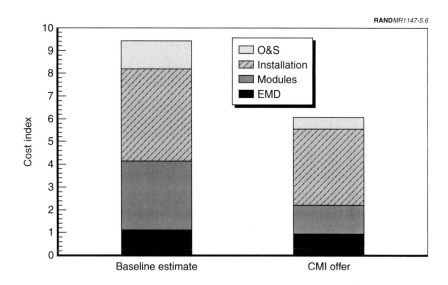

**Figure 5.6—Baseline vs. CMI Life-Cycle Cost Estimates for
Fighter Avionics Upgrade: Module and Installation Costs**

insert higher-performance, lower-cost parts that would constantly
reduce costs while the price remained fixed. The contractor estimated that all the important electronic parts within a module could
be replaced three times during a 10-year support contract, because
the average market life-span of a commercial electronics technology
is about three years. This would provide the opportunity to greatly
reduce costs and increase profit during the later phases of the support contract.

For these reasons, the contractor insisted that long-term logistics
support plus configuration control and change authority were crucial
for the credibility of its CMI bid. While elements within the Air Force
found this CMI offer attractive, the high cost of module installation
and the competing demands of other programs led to a temporary
shelving of the project.

Although many remain skeptical about the CMI proposal discussed
above, growing numbers of observers recognize that the problem of
rapid obsolescence and increasing unavailability of Mil-Spec avion-

ics parts is becoming severe, and might lead to the necessity of greater contractor configuration control and long-term contractor support contracts. The problem is caused by the phenomenon of DMS, Out of Production Parts (OPP), and rapid turnover and obsolescence of commercial technology. Many observers suggest that some form of continuous insertion, combined with contractor configuration control, will be necessary to maintain weapon systems over long periods of time, regardless of any DoD decision or strategy regarding CMI.

The phenomenon of DMS can be described as the rapid shrinking of the lower-tier vendor industrial base that historically provided low-volume, high-performance Mil-Spec and custom-designed parts for military applications. This shrinkage has been caused by the continuing decline in military demand, which itself has been brought about by a combination of reductions in military R&D and procurement budgets, the new emphasis on insertion of commercial-grade parts, and, most important, the vast increase in the relative size of the commercial electronics market compared with the military market.

A commonly repeated example of the latter phenomenon is the dramatic decline in military share of the IC market. As shown in Figure 5.7, the military share of the IC market declined from over 15 percent in 1975 to under 2 percent in 1995. By 1997, military demand accounted for less than 1 percent of global demand for ICs.[7] The military customer now has relatively little leverage in this market, particularly in very-low-volume, high-complexity, custom-designed ICs.

The end result of these phenomena is that, whether or not the services and contractors view CMI as a desirable strategy, the insertion of commercial-grade parts will increasingly be the only option available to military avionics developers. This would be viewed as a entirely positive development by advocates of CMI except for (at least) two problems that it causes: Increasing difficulties in supporting existing all–Mil-Spec "legacy" systems because of the OPP problem and

---

[7]Still, at around $1 billion, the military market is not insignificant.

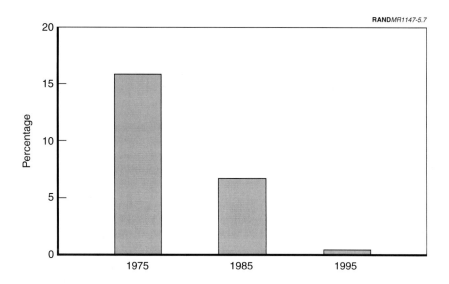

Figure 5.7—Declining Military IC Market Share

premature obsolescence and OPP for developers of new systems because of rapid commercial technology turnover.

The DMS phenomenon and the OPP problem for legacy systems are
well illustrated by the ALQ-99 Obsolescence Study conducted for the
U.S. Navy by AIL. Upgraded versions of the ALQ-99, an EW system
originally developed in the 1960s, are deployed on U.S. Air Force EF-
111As and U.S. Navy EA-6Bs. The study found that at least 175 parts
out of about 1000 examined were no longer manufactured by any
vendor. This finding in fact understated the severity of the problem
because it did not include out-of-production parts that the Navy had
stockpiled in large numbers prior to the termination of production.
The study also found that at least 15 major manufacturers of key
parts, mostly RF/microwave specialty components, no longer
existed. Finally, the study encountered severe difficulties in finding
appropriate commercial or alternative Mil-Spec parts that could be
substituted for the out-of-production parts, mostly because of
problems with fit (footprint). In short, inserting substitute commercial parts would often require major redesign and new R&D.

Perhaps more important for our research, studies conducted by the F-22 System Program Office (SPO) and other organizations show that the OPP problem is equally serious for new systems now under development, even if they are designed from the beginning to incorporate commercial-grade parts. Studies by the Semiconductor Industry Association (SIA) and others have determined that the average life-cycle of a commercial-grade IC or other complex electrical part is from two to five years (SIA, 1996). As late as the 1980s, life-cycles for parts averaged five to 12 years. Using the example of several generations of standard Intel microprocessor chips, Figure 5.8 illustrates the problems posed by insertion of commercial-grade parts with three-year life-cycles into the typical large system acquisition program that has an R&D schedule of 10 years and an inventory life of 30 years or more. As the figure shows, the chip that is commercially available at the completion of system R&D may be three or four generations beyond the original chip designed into the system years earlier, and may not be backward compatible.

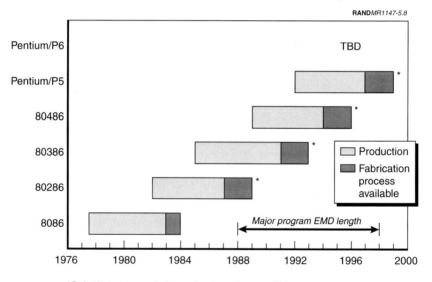

*Substitutes are not always backward compatible.

Figure 5.8—Out-of-Production Parts Problem: COTS Intel Example

This problem has been evident for some time on the F-22 fighter program, which is still in its EMD phase. Hundreds of parts have been identified that are already unavailable or will terminate production in the next five years. Scores of parts had become unavailable before the first flight of the first EMD prototype in September 1997. Many of these parts are not exotic Mil-Spec devices but rather commercial-grade off-the-shelf parts. In some cases, the parts become completely unavailable. In others, small after-market vendors buy the technical data packages and produce the parts in small custom batches, but at very high prices. The OPP problem means that avionics modules and components on the F-22 may have to be re-designed several times during and after production to incorporate new types of parts to keep the avionics systems operable.

In the view of some observers, even if commercial-grade off-the-shelf parts are inserted into avionics system to the maximum extent feasible, the problems posed by OPP confront the government with basically three stark choices:

- Buy and maintain large stockpiles of parts in quantities necessary to support the system throughout its entire life-cycle.

- Pay high enough prices to encourage an adequate number of specialty after-market vendors to continue manufacturing small quantities of a wide variety of obsolete parts.

- Grant configuration control and change authority to the integrator contractor, along with long-term fixed-price support contracts.

Many observers believe that the third option would be the most cost-effective. In principle, it would provide financial incentives to encourage continuous insertion of newer, less-costly, and more-capable technology into avionics systems, while at the same time resolving the OPP problem. Why then should the government not implement a similar policy of continuous insertion through its own depot infrastructure? There are no technical barriers to such an approach, but there are at least two reasons why it may not be cost-effective:

- First, the contractor's ability to provide the lowest-cost bid possible for the R&D phase may depend on the assumption of contractor logistics support with configuration control, because it is based on a calculation of potential dramatic cost reductions in parts and technology in future years. Some contractors claim that as much as 50 percent of the cost difference between CMI bids and normal bids is based on this projection.

- Second, although the government depot may have good intentions, it may not have adequate incentives to implement a policy of continuous technology insertion during the support phase of the life-cycle. A policy of continuous insertion implies the necessity for new R&D, which means new development costs and new technological risks. A private contractor has the financial incentives to take on such costs and risks because they can result in more-reliable, lower-cost modules. With a fixed-price support contract, this means higher profits. The government depot has no such incentives to motivate taking the risk of technological failure and increased R&D expenditures.

Whether such a program could be effectively structured in the real world with the appropriate incentives for the contractor is another question. Although careful assessment of this question is a central focus of our ongoing research, little light is shed on the issue in the case studies we have examined to date. But our case studies do suggest that the insertion of commercial-grade parts and technologies has a cost-savings potential of approximately 20 to 50 percent on the level of digital technology avionics modules. They also suggest that, because of the DMS problem, the use of commercial-grade parts will increase dramatically whether or not a comprehensive CMI strategy is already in place that, among other issues, deals with parts obsolescence, OPP, contractor configuration control, and contractor logistics support.

Chapter Six

LESSONS FROM THE COMMERCIAL AEROSPACE MARKET

The case study evidence presented in Chapters Three through Five of this report suggests that, in the critical area of radar-related microwave and millimeter-wave technologies, the dual-use promise of CMI and related acquisition reform measures is real. The technological breadth and depth necessary to support a comprehensive CMI strategy appears to be emerging, and there is some evidence to suggest that commercial-grade parts and components can be successfully inserted into RF/microwave military avionics systems without degrading system performance.[1]

But the evidence also indicates that to take full advantage of the overlap between commercial and military products and processes, DoD and its contractors must adopt a more commercial-like approach to weapon system acquisition. For example, it may be necessary for DoD to grant configuration control and change authority to contractors, at least at the subsystem and parts level, throughout the life of the system. This practice, widespread in the commercial world, will require that DoD and contractors alike be fully aware of the performance, schedule, and cost priorities for different weapon systems.

DoD may also benefit from adopting other mechanisms commercial firms—and particularly the "best" commercial firms—have developed to minimize the risks of inadequate product performance and

---

[1]Legitimate concerns remain about the long-term reliability and durability of commercial-grade parts and components.

excessively high costs. Many of these mechanisms are now being tested in DoD pilot programs; others have already been widely embraced. For example, to reduce the chance that weapon systems will fail to meet high-priority performance and cost objectives, Integrated Product Teams (IPTs) are being implemented throughout the DoD acquisition process to the maximum extent practicable. "Best value," as opposed to "lowest bid" source selections are now DoD policy. And many military specifications and standards documents have been either inactivated or cancelled. Other mechanisms appear to be underused. For example, DoD rarely engages in R&D risk-sharing arrangements with its contractors, although this practice could discourage some of the program cost escalations that have plagued it in the past. In addition, in large part because of the continued prevalence of cost-type R&D contracts, stringent cost-reporting requirements still apply to "noncommercial" items; these have a considerable cost impact on most weapons programs.

Nevertheless, there are legitimate reasons for caution. Before throwing out such important policy instruments as TINA, CAS, and various audit and oversight requirements, for example, DoD should evaluate the alternative mechanisms commercial businesses use to control costs and deter fraud. Similarly, before relinquishing ownership and control of technical data, DoD should consider the long-term implications for the reliability and supportability of its systems.

More broadly, some of the questions that DoD must consider before choosing to adopt wholesale a commercial-like approach to acquisition include:

1. To what extent can system cost be reduced without sacrificing performance?

2. Will qualified contractors be willing to absorb more of the market and technical risk associated with new aircraft system and subsystem development?

3. Can DoD promote and maintain adequate levels of competition in the absence of heavy regulation?

4. Can DoD ensure the supportability and maintainability of systems over time if contractors retain configuration control and change authority?

In this chapter, we examine ways in which U.S. participants in the market for large transport aircraft have approached similar questions, highlighting their relevance to DoD. Combining information from a variety of sources, including interviews with industry professionals, on-site plant visits, and various published and unpublished materials on commercial practices and processes, we draw an analogy between commercial and military aircraft manufacturers and between commercial airlines and DoD.

We choose the large transport aircraft market for our case study for three reasons. First, the commercial airliner market has many structural similarities to the market for military aircraft. These similarities suggest that the same sorts of strategies may be effective in both markets. Second, in the 1980s and 1990s, a fundamental movement toward price-based competition among U.S. airlines, airframe integrators, and aircraft system and subsystem suppliers has forced all firms in the industry to rethink the way they do business. Their responses to the pressures introduced by airline deregulation and to the emergence of strong foreign competition provide useful lessons both for DoD and military contractors who must learn how to operate in a world that is now severely cost-constrained.[2] Third, because many aircraft manufacturers operate in both commercial and military markets, their responses to changes in the commercial sphere may tell us a great deal about how they will respond to changes in the military sphere.

We suggest that, as DoD's approach to airborne weapon systems acquisition begins more closely to resemble the approach used by commercial airlines to purchase airliners, military contractors will respond by adopting many of the same strategies now prevalent among commercial aircraft manufacturers. In sum, DoD may expect to see an acceleration of the following trends:

---

[2]Although the number of firms in the commercial aircraft industry has declined, competition as measured by price pressure has increased with the entry of Airbus, a point that is discussed in more detail below. The situation is similar to that of the U.S. auto industry, which underwent a period of consolidation at the same time as—and partly because of—Japanese entry into the market in the 1970s. Unless otherwise noted, our use of the term "competition" refers to the degree of downward pressure on prices, not the number of competitors.

1. Greater emphasis by contractors on lowering the cost of purchasing and operating military aircraft as opposed to improving their performance characteristics.

2. Greater market and technical risk-sharing between prime contractors and suppliers of military aircraft systems, subsystems, parts, and components.

3. More intense competition accompanied by increased industry consolidation and greater foreign participation at all levels of the industry supply chain.

4. Greater integration of military aircraft R&D with maintenance, repair, and overhaul activities.

STRUCTURAL SIMILARITIES: MILITARY AND COMMERCIAL AIRCRAFT MARKETS

As discussed in Chapter Two, the structural characteristics of the markets in which transactors operate in large part determine the types of risks that they face. For example, in markets where there are many possible buyers for a product, market risk is relatively small. Stated another way, the risk that a firm will have to write off an investment because no one buys the product is higher in markets where potential buyers are few. Similarly, in markets involving new and complex technologies, technical risk is higher than in markets where technologies are well known. That is, the risk of being forced to write off an investment because the product fails to meet buyers' desired performance capabilities is higher in markets where technologies are still being developed.

Table 6.1 provides a generalized characterization of three types of markets: military aircraft markets, mass commercial product markets, and commercial aircraft markets. As the table makes clear, military aircraft markets do not look much like mass commercial product markets such as those for consumer electronics, automobiles, or microprocessors. Military aircraft markets generally have small production runs; potential buyers are few and their requirements specific; technologies tend to be expensive and untried; performance requirements at initial purchase and beyond are stringent; and the tolerance for performance variability is extremely

Table 6.1

Structural Characteristics of Commercial and Military Markets

Characteristic	Military Aircraft	Type of Market Mass Product	Commercial Aircraft
Output Quantity			
Total production	Small	Large	Small
Rate of production	Small	Large	Small
Nature of Demand			
Number of buyers	One buyer	Many buyers	Few buyers[a]
Who defines product?	Buyer	Seller	Seller, with significant buyer input
Demand stability	Highly uncertain	Fairly stable	Cyclical
Nature of Technology Technological			
challenge	Very high	Generally low	High
Learning effects	Important throughout production run	Modest at mature production	Important throughout production run
Performance and Service Requirements			
Level of performance	Stringent	Nonstringent	Stringent
Variability of performance	Nontolerant	Tolerant	Nontolerant
After-market service	Extensive	Limited	Extensive

NOTE: The description of markets is highly generalized. Many exceptions exist.

[a]Although there are well over 100 airlines, a few of the largest effectively determine the success of a new aircraft model.

low. To a large extent, these characteristics are inherent to the nature of the demand for military aircraft, and will not be affected by the introduction of commercial approaches to acquisition. To identify the commercial business practices that are potentially most useful to DoD, therefore, we should look to markets whose structural features most closely resemble military markets.

As Table 6.1 suggests, the military aircraft market has few features in common with commercial mass markets, but has several features in common with commercial aircraft markets. For example, as with military aircraft, low rates of production for and small total outputs of commercial airframes, aircraft systems, and subsystems result in high per-unit costs. Further, commercial airliner production is characterized by strong learning effects. Average production costs decline steeply over time in part because design changes may continue to be introduced well after the start of full-scale production.[3] Although commercial aircraft are not made-to-order in the same way as military aircraft, manufacturers must still be highly responsive to the needs of airlines, who are able to influence many performance aspects of the aircraft they buy. Finally, firms in the commercial aircraft industry employ stringent measures to limit the performance variability of the aircraft they build, and have after-sales service and maintenance requirements comparable to those found in the military aircraft sector.

There are, however, at least two important ways in which the market for military aircraft differs from the commercial aircraft market. First, unlike DoD, airlines rarely request that new airliners incorporate dramatic technology innovations. Instead, their approach to improving aircraft performance tends to be incremental, with a heavy emphasis on cost. Thus, although commercial transports are technologically complex by the standards of most consumer products, they generally involve less technical risk than military aircraft.[4] Because commercial airframe integrators and their suppliers can be fairly certain of meeting the performance requirements called for by airlines, they are willing to invest huge sums in new aircraft development. To get military contractors to do likewise, it may be necessary for DoD to lower the technical risk of military aircraft development by taking a more incremental approach to the introduction of new technologies.

---

[3]For this reason, early studies of production learning curves such as Berghell (1944) often focused on the aircraft industry. In recent years, investments in computer-aided design and manufacturing programs have been largely motivated by a desire to reduce expensive last-minute design changes.

[4]On the commercial side, possible exceptions include Pan American World Airways' role in encouraging the development of the Boeing 747 and the supersonic transport, discussed at greater length below. On the military side, new military transports are generally less technologically challenging than new fighter aircraft. An additional explanation for airlines' generally more conservative approach to innovation is their desire to maintain commonality of equipment—and thus lower maintenance costs—on their fleets.

The second way in which the military aircraft market differs from the commercial market concerns the unpredictability of demand even under conditions of no technological uncertainty. In the commercial sector, multiple independent airlines represent possible buyers for commercial aircraft. If one airline chooses not to buy a newly developed aircraft, the manufacturer has—at least in theory—plenty of other opportunities to convince others of the advantages of purchasing the plane. In contrast, defense contractors must rely on DoD and on the often variable political factors that influence both the level and composition of DoD's budget to make a sale. Defense contractors' inability to diversify their customer base for military aircraft significantly raises the market risk they face relative to firms operating in the commercial world.

In this respect at least, the difference between the commercial and military worlds is not quite as big as it may at first seem. One reason is that DoD does not act entirely as a single buyer: The different military services have different requirements for aircraft, so that failure to sell to one does not preclude success with another. Foreign military sales, although restricted, provide another possible market outlet. A second reason is that airlines do not act entirely as independent buyers: Although there are many commercial airlines, a few very large carriers are responsible for the majority of aircraft sales.[5] As we discuss in more detail below, it is often in the interest of these airlines to cooperate with each other both with respect to defining performance requirements and actually choosing which new aircraft to buy. Further, because airline demand for equipment is heavily influenced by movements in broad macroeconomic variables such as energy prices and economic growth, these carriers' requirements tend to be quite similar in terms of what they want and when they want it. This means that shifts in the demand for different types and quantities of aircraft tend to be synchronous, effectively mimicking the behavior of a single buyer. Market-risk-management mechanisms developed by the commercial aircraft industry, therefore, may still have relevance to the single-buyer military world.

---

[5]In the U.S. domestic market, 10 airlines accounted for 93 percent of revenue passenger miles in 1994 (Kaplan, 1995, p. 151). Because U.S. trunk airlines are among the only carriers able to make single orders of 20 or more aircraft, they are critical to the successful launching of new airliners (Zhang, 1996, p. 2).

THE RISE OF "MUST COST"

Not so long ago, the U.S. commercial aircraft industry exhibited many of the features still common to the military aircraft industry: widespread use of simple cost-plus R&D contracts, substantial government-financed R&D applicable to commercial air transport, long development cycles, and limited price-based competition.[6] In the past two decades, however, two events have combined to change the industry dramatically: the passage of the Airline Deregulation Act of 1978 and the emergence of strong foreign competition for U.S. commercial airframe integrators and their U.S. system and subsystem suppliers.[7] The transition of the U.S. airline industry, and by extension, the U.S. commercial aircraft industry, from an environment dominated by government regulation to an environment of free and fierce competition provides a useful model for DoD and its military contractors, who are undergoing a similar and no less profound transition as a consequence of severe defense budget constraints.

The Airline Deregulation Act of 1978 forced a profound shift in the way that U.S. airlines (and to a lesser extent airlines worldwide) operate.[8] Prior to 1978, the regulated fare structure set up by the Civil Aeronautics Board (CAB) required all airlines to charge the same fare for the same service.[9] Furthermore, entry into the air transport in-

---

[6]We use "simple cost-plus" contracts as a shorthand for contracts in which the buyer assumes all the risk associated with cost overruns. The percentage of total U.S. aerospace industry R&D financed by the private sector was just 11 percent in 1960, reaching 22 percent by 1975. Since approximately 1985 it has hovered around 34 percent (AIA, various issues). These calculations include defense R&D as well R&D related to space vehicles, so they overstate the government's contribution to commercial aircraft development. However, in basic research where multiple applications are possible, the overstatement is not so severe.

[7]Less important, but still significant, has been the sharp decline in U.S. government-financed aerospace R&D over the period, which in real terms fell by over 50 percent between 1987 and 1996 (AIA, 1998/99).

[8]Deregulation of air cargo service, which occurred one year earlier, also had an impact on the industry by encouraging the expansion of highly competitive all-cargo carriers such as FedEx and United Parcel Service (UPS). FedEx and UPS have been launch customers for all-cargo versions of Boeing, McDonnell Douglas, and Airbus transports as well as smaller aircraft.

[9]Not all U.S. airlines were subject to CAB regulations: Intrastate carriers, such as those operating entirely within California, were regulated by state public utilities commissions. See Jordan (1970).

dustry was strictly controlled. Because competition based on price was not allowed, carriers competed largely on the basis of other performance characteristics, such as their route structures and the quality of their in-flight service. Equipment cost-minimization was important but not critical because airlines were often able to pass higher equipment costs onto consumers by petitioning for higher fares from the CAB.[10] For example, as jet engine technology began to be commercialized in the late 1950s, the CAB allowed airlines to impose a $10 per ticket surcharge for travel on jet aircraft (Jordan, 1970).

When the barrier provided by protective regulation was abruptly removed in 1978, a flood of new low-cost airlines entered the air transport industry. Intense competition from these new entrants forced several of the established, high-cost airlines into bankruptcy.[11] Those that survived did so in large part by tightening their control over costs. Because of the difficulty of reducing labor costs, which account for a major share of the total costs associated with running an airline, reducing costs associated with the purchase and operation of aircraft and aircraft equipment became a major priority for U.S. airlines.

U.S. carriers' ability to put downward pressure on the cost of buying new aircraft was greatly aided by the emergence of strong overseas competitors to U.S. builders of commercial aircraft. In particular, by the late 1970s the European consortium Airbus Industrie had become a viable alternative to the U.S. commercial airframe integrators—Boeing, Lockheed, and McDonnell Douglas. Offering a product that was extremely price-competitive, by 1979 Airbus had won 26 percent of the market by volume for twin-engine wide-bodied jets. By the end of 1997, Airbus accounted for almost 50 percent of new orders for airliners, or 43 percent by value (Sutton,

---

[10]For example, according to Mann (1982), a 1980 survey conducted by the U.S. International Trade Commission found that price ranked fifth among 15 criteria cited by commuter airlines as important to their aircraft purchasing decisions. The first four criteria, ranked in descending order, were passenger capacity, fuel efficiency, quality, and technology.

[11]Between 1978 and 1985, more than 250 new airlines tried to enter the U.S. market, while 260 airlines went bankrupt (Zhang, 1996). Arpey (1995) estimates that bankrupt carriers controlled nearly 20 percent of U.S. airline capacity between 1990 and 1994.

1998).[12] Airbus' success also helped its own preferred system and subsystem suppliers, non-European as well as European, to grow strong. Although many of these suppliers are relatively new to world markets, they now challenge the dominance of the U.S.-based industry leaders.[13]

The combined effect of airline deregulation and increased foreign entry has been to encourage intense competition between airlines and between rival teams consisting of the airframe integrators and their preferred aircraft system, subsystem, and parts suppliers. At all levels of the industry, a new hard-line costing approach, called "must cost," is spurring firms to adopt radical cost-cutting measures.[14] In an environment where failure to achieve price and performance targets incurs enormous financial penalties—and ultimately loss of contracts—suppliers at every level have scrambled to find innovative ways to get their costs down without sacrificing quality.

Under the "must cost" approach to buying aircraft, the airframe integrator first conducts market research to determine potential customer requirements for a new airplane, collecting information about the price per plane as well as other performance- and operating-cost-related objectives desired by airlines. Using the price suggested by the airlines as a guide, the airframe integrator chooses a price and profit target for the finished aircraft, and by a combination of past experience and analysis of technical trends, determines the cost share of each major aircraft system. In consultation with prospective suppliers, the integrator then allocates rigorous price targets based

---

[12]These shares can change significantly depending on whether the measure is new orders, confirmed orders, or aircraft delivered. In 1998, for example, Airbus' share of new aircraft delivered was just 25.5 percent by value, in part reflecting Boeing's acquisition of McDonnell Douglas in 1997 (Flanigan, 1999, p. C-1).

[13]The European avionics firms that form Sextant Avionique, for example, had previously addressed only national military needs. Sextant is a preferred supplier for most Airbus aircraft and competes head-on with U.S. avionics suppliers. See Charles and Ghobrial (1995, p. 607).

[14]Although "must cost" does not mean that price targets are always binding—firms can and do sometimes exceed them—the term illustrates the new emphasis on cost that is found throughout the industry. See, for example, the discussions in Wilson (1996) describing the structure of "must cost" and its importance to McDonnell Douglas. Boeing's term for the "must cost" concept is "market-driven target costing" (Schwendeman, 1997); a description of the "must cost" process at work during development of the Boeing 777 is presented in Sabbagh (1996).

on those shares. In the case of Boeing, for example, this approach to aircraft development differs significantly from Boeing's previous linear approach of design, followed by engineering, and finally cost-estimating.[15] Figure 6.1 illustrates the new approach for Boeing's 757-300.

Figure 6.2 illustrates how "must cost" pricing is passed down the supply chain, using a stylized representation of the cost structure for

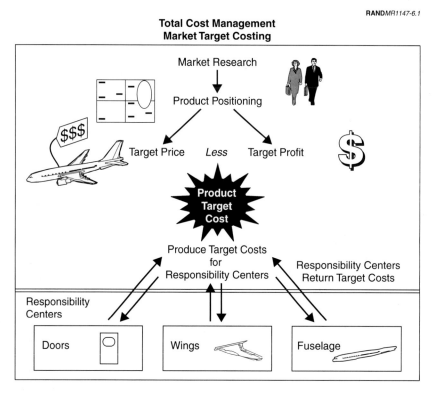

SOURCE: Schwendeman (1997).

Figure 6.1—Market Target Costing for the Boeing 757-300

---

[15]As described by David Schwendeman of the Finance and Business Management team within the Boeing Commercial Airplane Group (Schwendeman, 1997).

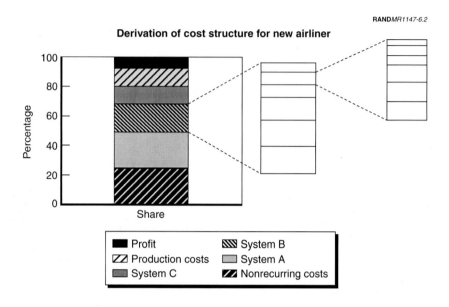

Figure 6.2—"Must Cost" Environment: Airframe Integrator's Cost Structure for New Aircraft Production

a new airliner. In this example, the price that the airframe integrator expects to receive from the airline is allocated among six major categories: nonrecurring costs such as those associated with EMD and program management; prices paid to suppliers for aircraft systems A, B, and C; production costs; and profits earned by the airframe integrator. As illustrated for system B, a similar breakdown exists for each aircraft system and subsystem. Each box represents a target cost share for the prime, system, or subsystem integrator.

The major difference between the "must cost" approach and that taken in the preregulation environment is that, under "must cost," price targets and not costs tend to drive the size of each box. If, for example, the box representing the cost share for system A gets too big, the system A integrator will find it difficult to pass on the increase to the airframe integrator.[16] In most cases, its most likely

_____

[16]Even under the old regulated system limits on cost pass-through imposed bounds on the sizes of the boxes.

options are either to find a way to reduce costs or to take a hit on profits.[17] For very large cost overruns that cannot be sufficiently reduced, and where attempts to renegotiate the contract fail, the system integrator must either undertake a discretionary breach of contract or file for bankruptcy.

Because of the priority placed on meeting price targets, "must cost" can only function well if integrators allow suppliers to choose how to design, manufacture, and service their own particular system, subsystem, or part. Instead of detailed technical specifications, products are defined in terms of generic performance requirements, form-fit-and-function (FFF) specifications, and the Federal Aviation Administration (FAA) safety standards required for new aircraft certification. This approach gives suppliers sufficient flexibility to respond to unexpected cost increases as well as to introduce new and more-cost-effective technologies. It represents a significant change from the old system, whereby the airframe integrator provided detailed product designs to manufacturing subcontractors who bore little responsibility for the product's ultimate performance.

The shift in technical responsibility from integrator to supplier under "must cost" has been accompanied by a parallel shift in financial responsibility: Suppliers now pay an increasing share of the R&D and certification costs for new aircraft. The increased financial commitment required to do business has encouraged aircraft industry suppliers to adopt one of two strategies—to "bulk up" by merging or allying with suitable partners or to become small niche players. The "bulk-up" strategy has been widely adopted, resulting in a profusion of business partnerships of every sort at every level, from simple risk-sharing arrangements to full joint ventures to outright mergers.

---

[17]A recent example described in *Flight International* (1998, p. 13) is an agreement between FedEx and Israeli Aircraft Industries (IAI), a licensed manufacturer of the 150-seat Fokker F-27. FedEx agreed to buy 100 specialized cargo versions of the F-27 developed by IAI provided that the unit price on each aircraft did not exceed $10 million. When IAI found that it would be difficult to meet the price target set by FedEx, it tried to convince FedEx to pay slightly more for the plane. FedEx refused to change the terms of the deal, and IAI is now looking for additional customers for the plane in order to increase the production run and so reduce per-unit costs. The company is also looking for additional subcontractors willing to share IAI's financial risk in return for the opportunity to manufacture major sections of the airplane.

Thus, "must cost" has contributed to an extensive financial restructuring and consolidation of the U.S. commercial aircraft industry.

Foreign firms have also played a role in the industry's restructuring. In addition to their financial capital and market access advantages, foreign firms provide U.S. firms with cutting-edge technologies and a wide range of products. In the avionics market, for example, the two biggest firms, both U.S.-based, have sought out foreign partners: Honeywell has teamed up with the British firm Racal Avionics to become the leading producers of satellite communications equipment, whereas Rockwell-Collins has joined with Dassault Electronique of France on ground collision-avoidance systems.[18] Well-financed and technologically sophisticated foreign firms counterbalance the growing market power U.S. firms are achieving through widespread industry consolidation. The presence of such European heavyweights as Airbus Industrie and Sextant Avionique, for example, greatly reduces the chance that airlines will be victimized by price-gouging on the part of the U.S. market leaders.

Finally, under "must cost," system and subsystem suppliers have increasingly sought cradle-to-grave arrangements with the airlines, whereby they agree to design, develop, manufacture, and provide after-sale support for the final product. The most extreme manifestations of this trend are "power-by-the-hour" and "fly-by-the-hour" contracts for engines and other aircraft systems that are offered as packaged solutions to airlines by the original equipment manufacturers (OEMs). In these types of arrangements, OEMs agree to maintain and service their systems on a long-term basis for a fixed fee based on the number of hours actually flown.[19] Such arrangements benefit the airlines because the promise of a fixed-price maintenance contract gives OEMs an incentive to improve the reli-

---

[18]AlliedSignal, the former third member of the avionics "Big Three," announced its merger with Honeywell in June 1999. The combined company is to be called Honeywell (Honeywell, 7 June 1999).

[19]For example, according to the Canaan Group (May 1994), Continental Airlines was able to secure a five-year, no-escalation contract from its maintenance, repair, and overhaul service providers. One reviewer commented that, given the menu-like character of these fixed-price maintenance, repair, and overhaul (MRO) contracts, in many particulars they closely resemble time-and-materials contracts. Time-and-materials contracts are not allowed under FAR Part 12, which governs DoD procurement of commercial items.

ability of designs and so reduce life-cycle costs for the aircraft. They benefit the OEMs because long-term relationships with airlines stave off competition from independent after-market suppliers toward the end of a system's life.

The "must cost" pricing constraints imposed on the commercial aircraft industry by the airlines in many ways parallel the DoD budget constraints that military aircraft manufacturers now face. The commercial industry's response to "must cost"—and the factors that are driving it—thus may provide important lessons to DoD. In the sections that follow we discuss how this highly constrained commercial pricing environment has led to

- a shift in emphasis from the performance to the cost of commercial aircraft;

- a reallocation of risk between buyers and suppliers;

- a stronger role for foreign firms as both competitors and allies of U.S. commercial aircraft and aircraft equipment manufacturers; and

- an enlarged role for OEMs in the long-term support and maintenance of commercial airliners.

STRIKING THE BALANCE BETWEEN COST AND PERFORMANCE

The promise of CMI is that applying commercial approaches and technologies to military aircraft development will improve performance while at the same time dramatically lowering life-cycle costs. A major concern is that the performance of U.S. military aircraft (and weapon systems more generally) will be compromised as a result of the commercial-market emphasis on minimizing costs. The concern is twofold. First, for any given system, there is the question whether technologies developed for commercial applications will perform as well in military environments, either because military environments are too rigorous or because the performance variability of commercially derived items is too great. The question here is the extent to which commercial technologies can truly be "dual use." We address this primarily in the first part of this report.

The second concern about CMI is in some ways more profound than whether commercial technologies can be dual use. It pertains to the nature of commercial-world tradeoffs between cost and performance. Can commercial approaches predicated on cost-minimization produce the highly innovative, high-performance technologies embodied in U.S. military aircraft?[20] To shed light on this question, in this section we examine how the rise of "must cost" has affected the cost-versus-performance tradeoffs chosen by the manufacturers of large transport aircraft.

Prior to airline deregulation, low cost was not a highly weighted objective of commercial aircraft manufacturers. A more important objective was to build technically sophisticated aircraft that could outperform those built by industry competitors.[21] Giving greater emphasis to performance, rather than cost, was legitimate from a business point of view because, under a unified fare structure, airlines could usually be persuaded to buy a more expensive but technologically superior new aircraft if convinced that their competitors were going to do so too. Cost increases associated with improved performance could be passed on to the flying public via the CAB. Thus, although contracts between airlines and aircraft manufacturers were not set up formally as cost-plus, the effect on incentives was similar.

The case of the Boeing 747 provides a prominent example of the airlines' "follow the technology leader" approach. In the 1960s, the only major airline pushing for the development of the 747 was Pan American Airways (Pan Am), whose strong-minded chief executive had already become a legend in the commercial aircraft industry. According to other airline industry executives, many U.S. and foreign carriers only reluctantly bought 747s because they feared the potential marketing advantages the plane might give their competitors,

---

[20]As discussed in Chapter Two, some CMI advocates believe commercial approaches lead to performance that is superior to performance achieved through traditional military approaches. Other observers claim that, at least in the commercial aircraft industry, the emphasis is on reliability at the expense of complex features, leading to performance that is "good enough" rather than "superior."

[21]Some authors argue that the aircraft industry's traditional emphasis on performance over cost developed because engineers—rather than professional managers— have tended to dominate top management. See, for example, Irving (1993, p. 201).

especially Pan Am. They bought the 747 despite their concern that the plane's immense size would make it uneconomical to operate. Newhouse (1982, pp. 121–22) quotes the former chief executive of Trans World Airlines (TWA) as saying: "We were reluctant participants in the 747. But we couldn't afford to sit it out. Had we known that the DC-10 and L-1011 were coming along, we might well have sat it out." Similar stories are told about American, Delta, and Eastern Airlines.[22]

Another example of technology and marketing outweighing cost concerns was the U.S. effort in the 1960s to develop a supersonic transport (SST). Although most U.S. airlines were not strong supporters of the SST, none was willing to concede the potential advantages of super-fast flight to a rival. As with the 747, Pan Am had expressed keen interest in such a plane.[23] Further, because the British and French were moving ahead with their own supersonic airliner, the Concorde, it seemed certain at the time that European carriers would soon be flying Concordes. According to Heppenheimer (1995, p. 229), "Boeing and the FAA estimated that even if the SST were restricted to overwater [supersonic] flights it could still sell five hundred airplanes." With no restrictions, their sales estimate was closer to 1200 planes. Thus, despite a lack of enthusiasm on the part of airlines (as well as potential passengers), Boeing and the FAA embarked on a $1.3 billion development program confident that by the time the SST was ready to fly, the demand would be there.

In the end, environmental concerns forced the FAA to abandon its efforts to develop the SST. Because the government had been paying for roughly 90 percent of the SST's development, when the FAA dropped the program in 1971, so did Boeing. After spending more than $130 million of its own funds, Boeing was unwilling and unable to assume the remaining technical and market risks associated with such an ambitious development program. Boeing's decision was

---

[22]Heppenheimer (1995, p. 223) suggests that, with the 747, Pan Am had "coerce[d] the domestic airlines into buying equipment they didn't need and could barely afford." Rodgers (1996, p. 251) argues that Pan Am "always set the standard for the latest in air travel, and other airlines had to order the new airplane to keep up."

[23]In June 1963, the President of Pan Am, Juan Trippe, challenged the U.S. government to increase its commitment to the program by announcing that Pan Am would take options on the European-built Concorde (Heppenheimer, 1995, p. 207).

validated when the European threat proved to be a chimera: With the onset of the first oil crisis of 1973–1974, any demand there might have been for the highly fuel-inefficient Concorde disappeared. Of the 20 Concordes that were built, all were sold to the (then government-owned) British and French national airlines.

In today's highly competitive, deregulated environment it is possible that such inherently risky projects as the 747 and SST might never have reached full-scale development. Rodgers (1996, pp. 243–244) claims that neither Boeing nor Pan Am conducted detailed analyses of projected costs and revenues for the 747 program. Apparently, Boeing executives preferred to rely on their own informal judgments about the plane's prospects. In 1996, by way of contrast, market analyses convinced Boeing executives not to risk going ahead with an approximately $10 billion program to develop a "superjumbo" version of the 747 with seating for between 500 to 1000 passengers.[24] Their prudence was rewarded on Wall Street with an immediate jump of over 6 percent in Boeing's share price (*The Economist*, 1997). And in February 1998, Airbus decided to delay its planned 1999 launch of a similarly sized aircraft, the Airbus A3XX, by at least nine months. Some analysts question whether, as a result of both technical and marketing problems, the A3XX will also eventually be drastically downsized or even cancelled (Lane, 1998).[25]

Efforts to develop a modern-day SST are also in jeopardy. Although NASA is bearing the technical risk associated with development of the High-Speed Commercial Transport (HSCT), an aircraft designed to carry 300 passengers at over twice the speed of most modern jetliners, the program is supposed to be privately funded after it reaches its targeted 2002 "technology readiness" date. But the HSCT is expected to reach profitability only if airlines can charge on average a 30 to 40 percent surcharge over subsonic fares (Saounatsos, 1998). This profitability constraint makes it doubtful whether the privately

---

[24]Boeing has since indicated that it may still be interested in developing a 550-seat transport that would have a high degree of commonality with the 747-400 and 777. One very large common feature could include the engines that were developed for the 777 (Proctor, 26 April 1999).

[25]In December 1999, Airbus announced plans to "gauge the potential demand" for a new 650-seat airliner, but many experts still believe such a large plane will be economically unviable (*Los Angeles Times*, 9 December 1999).

funded second phase of the program will ever be implemented (Sweetman, 1996).[26] Market research suggests that the aircraft's intended market—primarily first- and business-class passengers—are not willing to pay much of a premium for supersonic service (Sweetman, 1996; Saounatsos, 1998).

Consumers' preferences for cheap fares above all else—except for air safety and the frequency and convenience of flights—are driving airlines to adopt new aircraft acquisition strategies that strongly emphasize cost.[27] Under "must cost," carriers generally are not willing to pay for technology innovations that improve the performance of aircraft or aircraft equipment unless they believe those improvements will contribute to their immediate bottom-line profitability.[28] In November 1998, high-ranking representatives from up to 40 world airlines and aircraft leasing companies, including the largest, met to petition Airbus and Boeing to build less-expensive "no frills" transports. The two airframe integrators were asked to estimate how much of a price reduction the airlines could expect if they agreed to buy basic versions of airliners with many current options removed.[29]

Despite these indications that performance innovations are taking a back seat to cost containment, it would be misleading to argue that "must cost" has stifled technological improvements in the aviation industry. In some ways it is quite the opposite: Many cost-saving product redesigns and technology applications might not have taken place without the pressure of "must cost." In fact, the "must cost"

---

[26]Environmental concerns relating to noise levels are also reportedly placing the program in jeopardy.

[27]Consumers of air travel, and especially business travel, often strongly prefer the convenience of nonstop flights. On low-density intercontinental routes, however, large long-range aircraft such as 747s have relatively high seat-mile costs. The response to the airlines' requirement for long-range aircraft with lower seat-mile costs has been the development of twin-engine, twin-aisle planes such as the Boeing 777.

[28]In one industry interview, it was suggested that the required payback period for investments in older aircraft retrofits can be as short as six months.

We distinguish here between technology innovations that are strictly performance-related (such as increasing an aircraft's speed or range) and innovations that reduce operating costs. Innovations designed to increase an aircraft's fuel efficiency or operational reliability fall into the latter category.

[29]Customized features now account for approximately 4 percent of an average aircraft's price (*AWST,* 1998).

pricing system has so successfully met the basic requirements of high quality at low cost, it has been widely adopted throughout the commercial aircraft industry. From builders of all-cargo freighters to builders of business jets there has been a willingness to enforce and accept rigorous price and quality targets. The results can be seen in new aircraft prices. Zhang (1996), for example, estimates that prices for new jet aircraft, adjusted for quality differences and relative to the prices in the economy overall, fell by more than 12 percent between 1978 (one year before airline deregulation) and 1990.[30]

By and large, the "must cost" approach has delivered what consumers seem to want: Between 1979 and 1994, the average fare per passenger mile fell by more than 8 percent in real terms on U.S. domestic flights. In western and southwestern states, where many low-cost carriers have entered the market, fares on some routes have fallen by as much as 20 percent.[31] Further, the total number of scheduled departures from U.S. airports has increased by more than 50 percent and flights have become more reliable as well as more convenient.[32] Airline safety records have also improved since deregulation: According to the FAA (1996, p. 3), "in the 16 years prior to deregulation, there was an average of one fatal accident for every 814,000 flights. By 1994—16 years after deregulation—that figure dropped to one for every 2 million flights."[33]

---

[30]Calculated relative to the gross domestic product (GDP) deflator. For aircraft as a whole, that is, including military aircraft as well as single-engine piston-powered aircraft and business jets, new aircraft prices rose quite strongly relative to prices for all other commodities over the same period. Unadjusted for quality improvements, the real price increase for all new aircraft was 43 percent between 1978 and 1990 but approximately zero between 1986 and 1998. Authors' calculations based on producer price data from the Bureau of Labor Statistics.

[31]Based on GAO (April 1996) estimates. On some routes, fares have increased since deregulation.

[32]One study found that, since airline deregulation, there has been a significant increase in the number of engine hours between major overhauls, but in-flight shutdowns have not increased (Kennet, 1993).

[33]Nevertheless, some critics believe that aircraft manufacturers have been allowed to sacrifice safety for reductions in cost. For example, the move from three to two pilots on long-haul aircraft and the FAA's decision to allow two-engine aircraft to fly transoceanic routes have both been sharply criticized. See, for example, Galipault (1991).

What consumers have not gotten for the most part are planes that are significantly faster or more comfortable.[34] Technology innovations have focused overwhelmingly on two goals: reducing airline operating costs and maintaining airline safety records in the face of huge increases in air traffic. Improved engine and airframe technologies employed on Boeing's 777-300, for example, allow it to fly from San Francisco to Tokyo for up to one-third less fuel and 40 percent lower maintenance costs than similarly sized early-model 747s (Boeing, 1996–1998a). The time it takes the 777-300 to reach its destination, however, is not significantly different from that of early 747s, and in several important ways there are fewer passenger amenities than on earlier planes.[35]

Airlines have also been cautious about adopting innovations to onboard passenger services, and even ground operations. For example, multichannel satellite communications (SATCOM) systems were first developed in the late 1980s, making it technically possible for airlines to offer passenger services such as in-air fax transmissions and computer modem hookups. Most airlines, however, have only recently installed the necessary passenger equipment. Apparently, they were waiting for SATCOM prices to fall far enough to justify the installation cost. And although the reduction of operations and maintenance costs is clearly important to airlines, Boeing's attempt early on to outfit the 777 with an electronic library system—a system of hyper-linked graphical presentations of maintenance manuals, diagnostic procedures, wiring diagrams, minimum equipment lists, and many other features—was frustrated by airlines' unwillingness to pay the approximately $1 to $2 million estimated cost of installation per shipset (Charles and Ghobrial, 1995). Instead, Boeing decided to provide the plane with a much less ambitious onboard maintenance computer, leaving the electronic library system as an option available from the developer.

---

[34]Admittedly, incremental increases in aircraft speeds lead to far more than incremental increases in costs as speeds approach the sound barrier.

[35]On most if not all airlines the average seat size is smaller on newer planes, and there is less leg room in coach class (Gordon, 1990, pp. 124–129). Most airlines have also replaced the 747's optional upper-deck cocktail lounge with business and first class seating. Passenger entertainment systems, on the other hand, are far more elaborate than they used to be.

In fact, even many of the safety innovations introduced by the airlines may have been developed in the expectation that FAA mandates would force the airlines to buy them, rather than in response to market demand. For example, the traffic alert and collision avoidance systems (TCAS) and windshear detection equipment—now standard on all U.S.-registered transport aircraft—have been mandated by the FAA. Similarly, increased air traffic density, especially in the crowded skies over Europe, is causing the FAA to consider whether to require more precise navigation equipment on large transport aircraft. Many airlines are replacing their old long-range navigation and instrument landing system receivers with technologically more advanced global positioning system (GPS) navigation receivers, but it may be because they anticipate a government mandate.[36] Further, many of the innovations introduced on commercial airliners were derived from technology development programs sponsored by the U.S. government. All three products mentioned above—commercial TCAS, windshear detection, and GPS—contain technologies developed with U.S. government support. In the case of GPS, subsidization of commercial users continues.[37]

Our observation is that the "must cost" approach is delivering safe, reliable aircraft to the airlines at extremely competitive prices. However, a budget-induced design conservatism may also be reducing both the size and scope of purely performance-related technological innovations in the commercial aircraft industry. This implies that commercial approaches predicated on cost-minimization will not produce the kind of innovative, high-performance military aircraft desired by DoD unless performance considerations are given a higher weight in decisionmaking than is usual in the commercial world. Even before deregulation, most of the truly big innovations in aviation, such as supersonic aircraft or

---

[36]See, for example, market forecasts by Frost and Sullivan (1996). GPS reduces navigational drift from approximately two miles an hour to one mile an hour. This improvement becomes extremely significant for longer transoceanic flights.

[37]The Navigation System Using Time and Ranging (NAVSTAR) satellites tracked by GPS receivers were developed and produced by the U.S. Air Force. They continue to be maintained by DoD. The National Aeronautical and Space Administration was the first to develop windshear detection technology; it then encouraged private manufacturers to commercialize (NASA, 1995). Other aircraft innovations developed in part with government money include the high-bypass turbofan jet engine.

high-bypass turbojet engines or even jet engines themselves, were financed in whole or in part by governments. With the move toward incrementalism introduced by "must cost," these types of innovations may be less likely to appear.

FROM COST-PLUS SUBCONTRACTING TO SHARED-RISK PRODUCTION

A second issue is how to get contractors to finance a greater share of military research and development. The current system, under which DoD pays for 100 percent of R&D costs, is becoming problematic in an era of tight budget constraints. As the number of firms that DoD can afford to support on any particular R&D program declines, competition on those programs is reduced. With less competition and no close customer-supplier relationship with DoD, contractors' incentive to control costs in the later phases of new aircraft programs is also reduced.

A key feature of a "textbook" approach to acquisition is that firms bear the risks—as well as reap the rewards—of their own product development. However, in most textbook-type product markets there are many possible buyers. Firms can reasonably suppose that, if they come up with an attractive product at an attractive price, sufficient buyers will be found to earn an acceptable rate of return on their investment. In contrast, to recoup its investment, a defense contractor must rely on a single buyer, DoD, whose demand depends on highly variable international military and political factors as well as domestic political ones. Further, the capital investment required for developing new weapon system platforms, and especially military aircraft, is substantially greater than for most commercial investments.[38]

It is not surprising, then, that defense contractors are reluctant to use their own money to develop new military aircraft. Nevertheless, as we argue above, in this respect the difference between the commercial and military aircraft worlds is not as large as it may at first appear. The experience of the commercial aircraft industry suggests

---

[38]An apparent exception to this general rule is the Airbus A3XX, for which the estimated $10 billion development cost is comparable to the estimated $15 billion development cost of the Joint Strike Fighter (JSF).

that DoD does have the ability to exert leverage in this area. Commercial airliners, too, are hugely expensive to develop, and commercial airlines are notoriously fickle customers. Yet, under "must cost," airlines contribute less than ever before to the financing of new aircraft development programs.[39] How do they do it? We next examine how financial restructuring of the commercial large transport aircraft industry, plus a strategy of greater cooperation with airline customers and preferred suppliers, has helped commercial prime integrators adapt to the post-deregulation "must cost" environment.

Despite producing over half the value of most large airliners, aircraft system and subsystem suppliers generally absorbed few of the financial risks of development prior to deregulation. Most contracts between airframers and their suppliers were simple cost-plus arrangements, with a significant proportion of the cost of all design changes covered by the integrator. To keep suppliers honest, integrators spent large sums tracking and documenting changes.[40] There were occasional exceptions: Douglas Aircraft, for example, financed the development of the DC-9 (which became the MD-80 after the merger with McDonnell) by persuading some 20 equipment and component manufacturers to share in its development costs.[41] However, Douglas proposed these risk-sharing arrangements out of financial necessity and did not continue with them in its next major development program, the DC-10.

Simple cost-plus–type arrangements between the airframe prime integrators and suppliers of aircraft systems, subsystems, and parts were possible because, as we noted above, contracts between airlines and airframe integrators were also effectively cost-plus. In addition, airlines shared some of the risk of development by placing substantial downpayments on their orders for new aircraft. In the case of the Boeing 747, for example, just two months after agreeing on basic de-

---

[39] Information about the financing arrangements between airlines and airframers is highly proprietary, and arrangements also vary from purchase to purchase. This statement is based on descriptions provided in Rodgers (1996, pp. 351–353), Newhouse (1982, pp. 37, pp. 54–56), Heppenheimer (1995, pp. 307–311), and Sabbagh (1996, pp. 50–54).

[40] Based on industry interviews.

[41] See Bilstein (1996, p. 190).

sign concepts such as shape and approximate size, Pan Am signed a contract promising to pay Boeing half the total purchase price for 25 aircraft—with the first payment to be made well before delivery of the first aircraft. This was perceived to be a large amount even at the time, as downpayments of up to 25 percent were the general rule. Further, the 1966 contract between Pan Am and Boeing to build the 747 included a cost-escalation clause that allowed the initial unit price paid by Pan Am ($18.7 million in 1966 dollars) to rise with increases in labor and wholesale costs over the term of the contract. This type of cost-indexing soon became a standard feature of most aircraft purchase agreements.[42]

But in the 1990s, when the design and development of a new airliner has become a multi-billion-dollar enterprise, contracts for new aircraft are more likely to contain cost *de*-escalation clauses whereby manufacturers promise to reduce aircraft prices as manufacturing efficiencies grow. Further, because of the financial problems that have plagued the airline industry since deregulation, manufacturers frequently find themselves extending credit to the airlines for new aircraft orders instead of receiving downpayments.[43] Airframe integrators can not afford to take on full technical and financial responsibility for new aircraft development now that their ability to pass on unexpected cost increases is extremely limited. This is especially the case in a "must cost" pricing environment, where integrators often guarantee their price, schedule, and performance targets.[44] Boeing's recent difficulties with commercial aircraft production, for example, which included a 20-day shutdown of the 747-400 assembly line in November 1997, are estimated to have cost the company at least $2.6 billion in penalties (Lane, 1997).

---

[42]The basic unit price Pan Am agreed to pay for the 747 in April 1966 ($18.7 million) seems low considering that the price now listed by Boeing for the 747-400 is $167.5–$187.0 million in 1999 dollars, or approximately $45–$50 million in 1966 dollars. We do not have information on the unit price finally paid by Pan Am for delivery of its 25 aircraft, but Newhouse (1982, p. 120) hints that increases in the prices of labor and materials raised it considerably. For list prices on Boeing commercial transport aircraft, see www.boeing.com/commercial/prices/index.htm

[43]For example, Rodgers (1996, pp. 431–434) describes how the "airline recession" of the early 1990s affected the financing and sales of the Boeing 777.

[44]These targets are usually quite specific. On the 777, for example, Boeing offered separate weight, drag, and fuel consumption guarantees to each of its airline customers (Sabbagh, 1996, p. 191).

Given the enormity of the sums needed to finance the development of a new airliner, plus large penalties for failure to deliver on guarantees, airframe integrators have found that simple cost-plus–type contract arrangements with suppliers are no longer viable. These types of contracts not only do not solve the integrators' financing problems, they also do not provide suppliers with powerful enough incentives to meet aggressive cost and schedule targets. So today, integrators are asking the suppliers of major aircraft systems to become risk-sharing partners in new aircraft development programs.[45] In turn, the major systems integrators are demanding that their own subsystem, parts, and components manufacturers accept greater financial and technological responsibility for their products. Financing capability now ranks with technical performance as a criterion for choosing suppliers. As a rule, all major aircraft industry suppliers now finance most of their own R&D unless a prime integrator (or more often an airline) requests them to incorporate highly specialized features into the product design.

The extent of recent risk-sharing arrangements is illustrated by the fact that not only research and development costs but also flight testing and certification costs are often borne by suppliers. These costs can be quite high—one hour of flight testing can cost as much as $50,000, while certifying a landing system can cost upward of $500,000.[46] For new engine certification, industry analysts estimate that the cost approaches $1,000,000; an engine upgrade certification may cost half as much.[47] Further, national regulatory agencies in other countries often require suppliers to fulfill their own certification procedures. The cost of certification, therefore, is a major variable input into a firm's decision whether to launch a new program or upgrade an existing system.

Why then are commercial aircraft system, subsystem, and parts suppliers willing to assume such enormous risks when they have never done so in the past? A simple but not entirely simplistic an-

---

[45]However, suppliers rarely put up sufficient capital to entitle them to an equity stake in the new aircraft.

[46]Landing system certification is particularly expensive because the FAA requires 100 problem-free landings at multiple airports in diverse weather conditions to certify an automatic landing system, and many rehearsals are usually required.

[47]Based on industry interviews.

swer is that it is the only game in town: With Airbus, Boeing, and (prior to its merger with Boeing), McDonnell Douglas all requiring suppliers to participate in risk-sharing arrangements, suppliers can either agree to put up their capital or get out of the business.

But the answer goes deeper than this. In any industry, competitive firms recognize the value of good suppliers. To convince suppliers to stay in the business yet assume more of the costs of development, including testing and certification, the airframe prime integrators and aircraft system integrators have taken two related actions. First, they have initiated much closer working relationships with their suppliers than was true in the past. Second, they are reducing the number of suppliers with whom they deal. In sum, commercial aircraft prime and system integrators are adopting many of the "best" commercial practices mentioned in Chapter Two.

Airframe prime integrators now work much more closely with their suppliers than they did in the past. They are also encouraging their suppliers to talk directly with the airline customers, both independently and as part of IPTs. Today, airlines, integrators, and suppliers often cooperate on new aircraft all the way from conceptual design to EMD. During the early stages of the 777 program, for example, Boeing set up 250 IPTs (called "design-build teams" by Boeing), each responsible for a section of the aircraft. Each IPT consisted of engineers and managers from system, subsystem, and parts suppliers, from the launch airline customers, and from Boeing itself.[48] The approach proved so successful that Boeing is using it on its next generation of 717, 737, 757, and 767 aircraft. Boeing is not alone in its approach: Before its merger with Boeing, McDonnell Douglas took a similar approach in the design and development of the MD-95, the now renamed B-717 (Boeing, 1996–1998b).

This closer relationship between system and subsystem suppliers, airframe prime integrators, and airlines necessarily relies on mutual trust to a greater extent than in the past. Suppliers must be convinced that mechanisms such as IPTs will reduce their technical and, especially, market risk, helping to ensure that they earn reasonable

---

[48]In the case of the 777, launch airlines were also invited to attend design-build team meetings. United Airlines, All Nippon Airways, and British Airways were among the carriers who participated in early design-build teams (Sabbagh, 1996).

returns on any program-specific R&D investments they now must make. For IPTs to work properly, prime integrators as well as suppliers must allow each others' engineers, as well as engineers from the airlines, to observe and even participate in sensitive corporate decisions regarding design and manufacture. A high degree of trust is necessary because all of the participants, including the airlines, gain tremendous insights into proprietary information about the nature and costs of aircraft development and production.

Closer relationships do not mean that relations between IPT members are always sympathetic, or even entirely cordial. With or without the IPT structure, the pressure put on suppliers to meet their cost and schedule obligations is enormous. Important long-time suppliers to both Boeing and GE Aircraft Engines, for example, have recently been heavily pressured to reduce costs and shorten cycle times—while at the same time being asked to dramatically reduce product defects.[49] The penalties for missing targets can range from financial slaps-on-the wrist to loss of contract. But when IPTs are in place, solving problems becomes a team effort: Suppliers are urged—if not required—to identify problems early so they can be shared with other members of the IPT. On the Boeing 777's IPTs, for example, each member of a team was required to sign-off on problem-solving decisions made jointly by the team (Sabbagh, 1996, p. 74). In fact, even without the formal mechanism of IPTs, integrators are helping their suppliers improve in areas such as process efficiency by regularly visiting their plants and working closely with managers and engineers. Suggestions from the integrators, although often critical, help to make suppliers more competitive in future contract bids.

Another step that airframe and system integrators have taken to make risk-sharing more attractive is to reduce the number of suppliers with whom they deal. In avionics, for example, airframers are moving toward the concept of selectable "supplier furnished equipment" (SFE) suppliers, allowing airlines to choose among a few suppliers whose products then become part of the standard aircraft package. The airlines like this concept because it offers them a choice of avionics equipment without requiring them to make inde-

---

[49]In one case the response from a key supplier was deemed to be inadequate by the prime; the prime sent its own engineers to the responsible supplier facility and reviewed all costs associated with that facility down to an extremely detailed level.

pendent contract arrangements with suppliers.[50] The airframers like it because it allows them to eliminate the many "buyer furnished equipment" (BFE) suppliers whose equipment used to be available to the airlines as an option.

SFE suppliers are generally not required to bid for contracts at each phase of the design and manufacturing process, but are expected to be highly responsive to the needs of the integrator as part of a longer-term partnership. This strategy reassures suppliers of the integrator's commitment, avoids costs associated with frequent recompetitions, and puts SFE suppliers in a position to influence the establishment of future industry standards and specifications for new equipment. This last point in itself provides a considerable benefit to SFEs.

To conclude, the ability of aircraft industry suppliers to assume more of the risks as well as more of the benefits associated with new aircraft development has been a key element in the industry's success in controlling costs. Two factors have helped to make it possible: closer ties among airlines, airframe integrators, and aircraft industry suppliers, and growing consolidation of the supplier base. More than ever before, commercial airframe integrators, aircraft system integrators, and their subsystem, parts, and components suppliers are seeking to form partnerships and alliances, both U.S. and foreign. To the extent that DoD can also take advantage of these trends, it should be able to persuade contractors to accept a greater share of the costs of military aircraft development.

CONSOLIDATION AND COMPETITION

A major risk associated with DoD's adoption of a commercial approach to military aircraft acquisition is that there will be insufficient competition to prevent price-gouging by contractors. Under the current system, DoD regulations limit the profits that contractors can earn on most military contracts through the use of simple cost-plus contracts plus profit caps. But because simple cost-plus contracts

---

[50]Under the old system, airlines who did not choose to accept the SFE avionics package could request "buyer furnished equipment" avionics. BFE avionics were offered as an option by the airframer, but airlines had to negotiate their own contracts with BFE suppliers.

also tend to reduce cost-cutting incentives, it is not clear that the total cost to DoD of acquiring aircraft is minimized by this strategy. The "best" practices strategy of encouraging suppliers to share the development risk for complex products promises to reduce DoD's acquisition costs by giving firms the right cost-minimization incentives. For it to work, however, incentives must be sufficient for firms to pass their cost savings on to DoD.

In recent years, the U.S. military aircraft industry has undergone widespread consolidation. On the face of it, this would appear to make CMI less attractive to DoD because fewer firms in the industry presumably mean less competition and therefore higher prices. But it may in fact be cost-effective for DoD to have fewer, healthier suppliers who are comfortable making the exceedingly expensive, capital-intensive investments required for designing and manufacturing military aircraft. And besides, because the recent spate of defense-related mergers and acquisitions is largely a response to defense budget cuts, the price implications of consolidation are not clear.

In this section, we examine the effects of airline deregulation and the rise of "must cost" on the structure of the commercial aircraft industry and on the prices of commercial large transport aircraft and aircraft equipment. The experience of the commercial aircraft market may be instructive to DoD because commercial airlines, who have long sought ways to encourage competition among U.S. airframe and aircraft system and subsystem suppliers, are also facing a wave of consolidation, especially at the lower tiers of the industry. This consolidating trend is in large part being driven by strong competition between the airframe prime integrators and their adherence to "must cost." One factor in keeping the industry competitive at all levels has been the growing participation of foreign firms, particularly with respect to generic equipment and parts.

The structure of the commercial aircraft industry has long been oligopolistic at its upper tiers—within each broad sector of the industry, such as airframes or engines or avionics, just two or three firms have traditionally dominated the market. But at the lower tiers, aircraft subsystem, parts, and components suppliers, there have typically been thousands of small firms. This is now changing: By some estimates, the number of firms producing commercial aircraft parts and components has fallen by a factor of three.[51]

Analysts propose several factors to explain this consolidating trend, including the need to "bulk up" in response to increased financial responsibility, the practice by system integrators of restricting contract awards to preferred providers, and suppliers' desire to increase profitability through the production of higher-value-added integrated systems. The first two of these factors suggest that "must cost" is playing an important role in the consolidation process. The ability to assume risk is now one of the primary qualities that airframe and aircraft system integrators look for when choosing suppliers, and this requirement is being passed down to the level of subsystems and even parts suppliers.[52] Integrators are also simply reducing the number of suppliers with whom they are willing to do business. Wilson (1996, p. 7), for example, quotes a former Douglas Aircraft executive as saying, "As we optimize our suppliers, we only will do business with certified suppliers. We want to increase the amount of business we do with high-performing suppliers. There will be some fallout in total numbers of suppliers used as a result of that."

To the extent that further consolidation of the commercial industry is being driven by the exigencies of "must cost," therefore, we should not expect to see anticompetitive behavior at the lower tiers of the industry putting upward pressure on the prices of airliners. Nevertheless, airlines face a highly concentrated industry. How then do they exert downward pressure on prices when there are relatively few suppliers?

One way has been to cooperate with each other on new aircraft acquisition. Although individual contract terms with manufacturers are a strictly kept secret, airlines generally work together to come up with common and compatible requirements for new aircraft. By doing so they not only hope to create a market large enough to take advantage of sharply declining average unit production costs, but also large enough to encourage rival manufacturers to compete for that

---

[51]Based on industry interviews.

[52]Aerospatiale, the French partner on Airbus, identifies a solid financial base, technological capability, and quick response time as the primary qualities it looks for when choosing a supplier/partner (Cook and Macrae, 1997).

market. When rival manufacturers propose aircraft designs with broadly similar performance characteristics, airlines may try to influence each others' ordering decisions. Since none wants to "split the market" and so increase per-unit costs, the timing of new aircraft orders is a highly strategic business decision.[53]

The importance of this last point is illustrated by a well-known case in which U.S. airlines did split the market.[54] During the late 1960s, the then "Big Four" U.S. carriers—American, United, TWA, and Eastern—all agreed on the need for a new three-engine widebody transport capable of flying nonstop across the United States. Lockheed and McDonnell Douglas came up with very similar designs, and their initial competing bids were very close. Failure to agree on which plane to order, despite last-minute personal calls and meetings by senior airline executives, resulted in the Lockheed L-1011 and the Douglas DC-10 development programs both going forward at the same time. This had ultimately disastrous implications for Lockheed's commercial aircraft operations.[55] The airlines suffered too: TWA, Eastern, and Delta, who bought the L-1011, were stuck with an aircraft that had a production run of only 250 units; American and United, who bought the DC-10, also were forced to bear much higher costs in the long run because of the divided market.[56]

The more highly competitive environment created by the deregulation of the airlines has probably discouraged airline executives from making personal efforts to coordinate their new aircraft acquisition strategies. However, airlines still cooperate through other forums, including those provided by manufacturers. Boeing, for example,

---

[53]Choosing an aircraft manufacturer has become an even more strategic decision for the airlines now that there are only two manufacturers of large transport aircraft. Airlines must balance their immediate need for the best airplane at the best price with their desire to avoid a monopoly situation in the future.

[54]This example draws heavily on Newhouse (1982, pp. 141–160), who provides a highly readable account of the role of the airlines in the battle between the L-1011 and the DC-10.

[55]Many analysts argue that McDonnell Douglas was also financially crippled by the competition with Lockheed but simply took longer to die.

[56]Although the airlines were able to cut highly favorable initial deals with both Lockheed and Douglas, the manufacturers' failure to achieve significant economies of scale on their aircraft benefited no one in the end.

convened representatives of eight large airlines—United, American, Delta, British Airways, Japan Airlines, All Nippon Airways, Qantas, and Cathay Pacific—for nearly a year's worth of meetings to discuss design plans for the 777. Many of their suggestions were incorporated into the plane, and by November 1998 six of the eight airlines— and many more besides—had placed orders for the 777. In two key respects—requirements and aircraft choice—the airlines effectively turned themselves into a single buyer.

A second way in which airlines have tried to exert downward pressure on prices has been to promote standardization and interoperability of aircraft equipment and parts. In this way, they hope to ensure that no one firm has monopoly control over any crucial aircraft system, subsystem, or component. In avionics, for example, the airlines jointly established Aeronautical Radio Inc. (now ARINC) to develop commercial aviation standards for air transport avionics equipment.[57] One of the stated goals of ARINC is to establish "industry-defined products that can be produced on a competitive basis by various suppliers." A second goal is to "enable airlines and other avionics users to achieve economies of scale in the procurement of avionics . . . through the standardization of avionics form, fit and function and definition of aviation communication systems" (ARINC, 1998). Although in principle ARINC standards make entry into the avionics industry much easier, thereby encouraging competition, there are two limitations to this approach.

As in most industries, it has always been difficult for late entrants to a new avionics technology to compete with the technology leader. A well-publicized antitrust suit involving rival Inertial Reference Systems (IRS) produced by Honeywell and Litton provides a recent example. According to Carley (1996), in the 1970s Honeywell developed a new and much improved IRS based on a ring-laser gyroscope. The Honeywell IRS soon caught on with the airlines, and Litton, the former IRS industry leader, was forced to develop a competing ring-laser gyroscope. Unfortunately for Litton, its efforts proved to be too little, too late. Few airlines bought the Litton IRS, even though it was

---

[57]Initially, their goal was to establish standard radio frequencies for aviation use.

ARINC-compatible with existing navigation systems.[58] Of the jets using laser gyro systems in 1993, Honeywell systems sat on 97 percent of Boeing planes, 100 percent of McDonnell Douglas planes, and 77 percent of Airbus planes. In 1993, two federal antitrust suits brought by Litton charged that Honeywell had unfairly wielded its monopoly power to discourage the airlines from buying Litton's IRS. Litton won both suits, but Honeywell retains its dominance of the market.

In avionics today it may be even harder to compete with the technology leader because the principle of federation, in which individual suppliers provide stand-alone systems that connect to other systems through FFF specifications, appears to be losing out to the principle of modular integration, in which multiple systems are controlled by one or more central processors.[59] Technical as well as economic reasons for this shift exist, and the two are related. Technically, advances in processor technology have allowed system integrators to consolidate related avionics systems into a single system. Economically, system consolidation is generating economies of scope and scale in marketing, R&D, and production. As a business strategy, therefore, the big avionics system integrators have been aggressively acquiring firms that supply relevant avionics subsystems and components. Independent system and subsystem suppliers are finding it increasingly difficult to compete against the comprehensive product lines offered by their larger competitors.

Commercial flight management systems (FMS), for example, coordinate several different types of navigation, communications, and instrumentation equipment into a single piece of equipment. At current prices, FMS are quite cost-effective, allowing airlines more flexibility and thus more efficiency in air routes and landing approaches. But the success of these systems has cut into the market for stand-alone products, including various sensors and displays. FMS systems are now standard equipment on current-generation large airliners and are becoming increasingly common on smaller business and commuter aircraft as well.

---

[58]One reason is that early versions of the Litton system were plagued by technical problems (Carley, 1996).

[59]The concept of integrated modular avionics was originally developed for the F-22.

Flight management systems are just the tip of the iceberg. On the Boeing 777, Honeywell's Airplane Information Management System (AIMS) allows even greater modular integration of displays, flight management, flight-deck communications, airplane condition monitoring, thrust management, central maintenance, and digital flight-data acquisition. This eliminates the need for separate racks of Line Replaceable Units (LRUs) for each avionics subsystem, thus saving significant weight, space, and power consumption on board.[60] Although VIA 2000, the successor to AIMS, has yet to establish a solid market, many observers believe that its modular technology represents the wave of the future (Nordwall, August 1995).[61]

In sum, in the past ARINC standards appear to have promoted competition between avionics suppliers. However, they may become less relevant in an era of increasingly integrated digital avionics, where many ARINC-standard LRUs are likely to be replaced by plug-in cards.[62] Further, whereas once airframe integrators regularly provided BFE avionics suppliers with designs based on particular ARINC specifications, under the new "must cost" approach to airliner development, the role of BFE suppliers has been greatly reduced. And as SFE system suppliers accept more design responsibility, ARINC also decreases in importance. The result may be to increase the technological dominance of certain avionics suppliers—and especially the major system integrators—even more firmly than was true in the past.

The final and probably most important way in which U.S. airlines have put pressure on airliner prices has been to buy the best product at the best price regardless of where it is produced. In particular, af-

---

[60]Honeywell (1997) estimates a weight savings of 510 lb and a volume savings of 104 MCUs (Modular Concept Units) for a typical widebody aircraft incorporating its VIA 2000 modular avionics, the follow-on to AIMS. For a narrowbody aircraft, the estimated weight and volume savings are 350 lb and 34 MCUs, respectively.

[61]"VIA" stands for Versatile Integrated Avionics. VIA 2000 has been selected as the avionics architecture for the Boeing 717's Advanced Flight Deck, but few of these planes have been sold so far (Honeywell, 7 September 1998).

[62]For example, it has been pointed out to us that Aircraft Communications Addressing and Reporting System (ACARS) units have always been designed to an ARINC FFF standard. In the AIMS cabinet on the 777, however, there has been no attempt to write an ARINC-type standard for the ACARS unit because the architecture is inherently proprietary.

ter nearly 10 years of ignoring Airbus—and continuing to buy their airliners exclusively from U.S. manufacturers—U.S. airlines finally decided that the European consortium was in the market to stay and that it had a good product. Although Airbus did not achieve its first U.S. sale until 1978, the year that the Airline Deregulation Act was passed, by the mid to late 1990s Airbus had established a significant market presence in the U.S. market.[63] In 1998, according to Airbus, "Eight out of eleven major airlines now operate or hold orders for Airbus Industrie aircraft. Customers placing some of the year's largest orders include United Airlines and US Airways."[64]

Both Boeing and Airbus—and prior to the merger with Boeing, McDonnell Douglas—have responded to the cost pressures introduced by the airlines by establishing international networks of suppliers. To some degree, the internationalization of their production lines may have been driven by concerns over foreign market access and the need for capital rather than comparative advantage. Greater foreign participation in the commercial aircraft industry does not necessarily mean it is becoming more efficient. The evidence suggests, however, that the pressure on prices is growing: For example, some economists estimate that Airbus' entry into the commercial airliner market may have lowered the average price of a long-range widebody airliner by more than 3 percent in the 1980s.[65]

In fact, some of the biggest competitive challenges to the primarily U.S.-based industry leaders are now coming either from foreign firms or from U.S.-foreign alliances. In some sectors of the industry, such as engines, this has long been the case. In other sectors, such as avionics and aerostructures, the growth of foreign competition is more recent.

---

[63]Airbus' first potential U.S. customer was Western Airlines, a California commuter airline, but the deal fell through. Eastern Airlines was the first U.S. airline to consummate a deal with Airbus, leasing four A300B4s in 1977. Eastern decided to purchase its leased Airbus planes in 1978.

[64]United Airlines and U.S. Airways are among the four largest airlines in the world, as measured by numbers of passengers carried (Airbus Industrie, 1999).

[65]Percentage change calculated relative to a base-run model simulation by Neven and Seabright (1995). According to some analysts, aggressive marketing duels between Boeing and Airbus in the mid-1990s have resulted in price cuts of 35 to 40 percent off list prices (Flanigan, 1999).

Foreign firms, both as partners of and as rivals to U.S. firms, have long been important producers of commercial aircraft engines. The British firm Rolls-Royce, which produced its first aircraft engines in 1914, actively competes in all segments of the aircraft engine industry. Its 1990 joint venture with Germany's BMW and 1995 acquisition of the U.S.-based Allison Engine Company have given Rolls-Royce a market presence in the corporate and short-haul regional jet markets as well as its traditional markets for medium- and long-haul airliners. Similarly, CFM International, the alliance between General Electric and Snecma of France, has produced a highly successful series of engines since 1974.[66] CFM engines power nearly 40 percent of all aircraft with a capacity of 100 passengers or more (CFM, 1998).

In contrast, significant foreign competition has only more recently emerged in the commercial avionics sector. For example, an alliance formed in 1998 between Smiths Industries (UK) and Sextant Avionique (France) is providing competition for Honeywell in the FMS market. The new Sextant/Smiths FMS is being offered on all new Airbus A319, A320, A321, A330, and A340 airliners.[67] Similarly, Rockwell Collins' preeminence in the traditional radio communications equipment sector is being challenged by an alliance between Honeywell and Racal Avionics (UK), the market leaders in the growing satellite communications field (Frost and Sullivan 1996). At the level of generic components and parts, foreign competition is becoming well-established.

In commercial-transport aerostructures manufacturing, significant rivals to Northrop Grumman (U.S.) include Alenia Aerospazio (Italy) and Short Brothers (UK), as well as the Japanese giants Mitsubishi Heavy Industries, Kawasaki Heavy Industries, and Fuji Heavy Industries. More and more, these firms are sharing the development risk with the airframe prime integrators. On the Boeing 757/767, for example, the three Japanese firms and Alenia (then Aeritalia) were risk-sharing major participants in development and production. Similarly, firms from Australia, Canada, Europe, Japan, and the

---

[66]Snecma stands for Société Nationale d'Etude et de Construction de Moteurs d'Aviation.

[67]Honeywell charges that the development of the Sextant/Smiths FMS for Airbus is being unfairly subsidized by the French government (Honeywell, 7 April 1998). Sextant is licensing the FMS technology from Smiths (Sextant, 1998).

United States provided sections of the 777 airframe to Boeing, with the Japanese firms alone designing and building roughly 20 percent of the airframe structure.[68]

In sum, the major subsectors of the commercial aircraft industry contain relatively few big players in the upper tiers, and the lower tiers of these sectors are consolidating. Nevertheless, there does not appear to have been a significant escalation in average prices for airliners. Airlines are helping to put downward pressure on the price of new airliners by pursuing the following strategies:

- Cooperating on new aircraft acquisition so as to avoid "splitting the market"

- Encouraging the standardization and interoperability of aircraft equipment and parts, but buying integrated systems where they are cost-effective

- Buying the best product at the best price regardless of where it is produced.

We believe that these strategies are also relevant to DoD, and that, despite recent consolidating trends within the military aircraft industry, there is still sufficient competition between the prime integrators to make CMI a viable strategy. DoD already coordinates weapon system acquisition for the military services, trying to ensure economies of scale whenever possible. This strategy should be continued, while remaining conscious that differences between the missions of the military services may not always allow for acquisition programs to be joint.

The nature of the economic and technical tradeoff between standardized federated systems and integrated modular systems for DoD is less clear. However, acquisition reform is allowing DoD greater latitude to make these choices on a system-by-system basis. In many cases, it may be in DoD's interest to ensure that Mil-Specs are not simply eliminated, but rather replaced by commercial standards and

---

[68]Originally, the three Japanese firms were also asked to become partners with Boeing on the 777, providing an equity stake of 25 percent. For political as well as economic reasons, the equity offer was later transformed into a fixed-price contract arrangement whereby the Japanese provided their own capital for development.

specifications. Issues concerning parts proliferation and support and maintenance are discussed at greater length in the section below.

Finally, DoD must also examine the tradeoff between its desire to keep sensitive military technologies out of the hands of potential enemies and the clear economic benefits of using dual-use commodities and technologies and pursuing best-value sourcing worldwide.[69] This issue pertains mostly to prime integrators and the suppliers of major systems. At lower tiers, it is probably already moot: Globalization of supply is almost as much of a fact for military aerospace as it is for the commercial aerospace world.[70] Further, we argue that relatively little information about military capability is revealed by commodity purchases because the security value-added lies in knowing how to integrate them into systems, as well as what the system requirements are.

PRESERVING SYSTEM SUPPORTABILITY

A sometimes underemphasized concern about adoption of a commercial-like approach to acquisition is the long-term supportability of military aircraft. One fear is that giving contractors increased control over aircraft and airborne weapon system configuration will encourage parts proliferation, creating severe inventory control problems for DoD. A deeper fear is that contractor configuration control could put DoD in a position where it is forced to outsource the long-term MRO of each system to its OEM. If standardized Mil-Specs parts and components are replaced by proprietary technologies, the argument goes, OEMs will be able to demand exorbitant prices for MRO services, effectively holding DoD hostage for future system support.

We discuss here how the competitive pressures introduced by airline deregulation have affected the market for commercial aircraft MRO

---

[69]In fact, DoD must take into account domestic political factors as well as international strategic factors when considering the foreign sourcing of military aircraft. The U.S. government, like most governments, is reluctant to spend domestic tax revenues on items produced in other countries.

[70]See the discussion in Lopez and Vadas (1991).

services.[71] Once again, we believe the commercial experience is relevant to DoD because, in circumstances similar to those DoD now faces, airlines appear to be reducing their costs significantly through judicious outsourcing of MRO services.

In parallel with the restructuring of the market for new commercial aircraft, profound changes have affected—and continue to affect— the commercial aircraft MRO market. These changes are taking place within an evolving relationship between airlines and suppliers. On the demand side, a new focus on core competencies and cost control has led airlines to seek to become, in the words of Robert Crandall, former Chief Executive Officer of American Airlines, "virtual airlines." A virtual airline is a company that, in the extreme, performs only its core task of carrying people and freight from point to point. A fully virtual airline would not own its planes—it would lease them—and it would outsource all activities, including overhaul and maintenance, other than flying people and cargo. Ideally, the airline would deal with only one supplier, which would hold all inventory and organize the entire supply chain for the airline.

On the supply side of the market, MRO activities used to be performed primarily by the airlines themselves. Even for systems under warranty, airlines would often be given in-house warranty authority, allowing them to fix systems themselves and bill the OEM for technician time and parts. Following the 1978 deregulation, however, many airlines just entering the industry did not have existing MRO facilities or spare parts inventories. The growth of these new low-cost airlines encouraged the entry of independent MRO suppliers who offered relatively low-cost services ranging from line maintenance to inventory control. In a responding effort to cut costs, several of the established airlines, such as Southwest and American, began to outsource more of their MRO activities. Still other airlines, such as United, took a different tack, expanding their own MRO operations by offering MRO services to third-party airlines.

Each type of supply structure had its own competitive advantages. As shown in Figure 6.3, OEMs were the primary providers of MRO

---

[71]We give a generic overview of the trends within the commercial MRO market. We recognize that there are distinct differences in the economics of MRO services for different aircraft systems (avionics versus engines, for example).

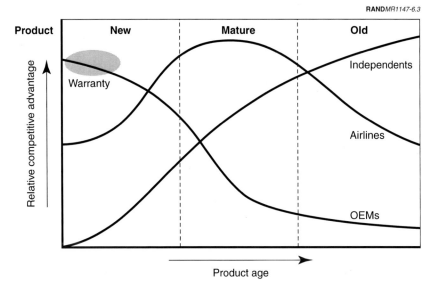

SOURCE: The Canaan Group (March 1996, p. 2).

Figure 6.3—Aftermarket Competitiveness

services during the early years of most systems because they offered warranties and controlled the spare parts pipeline.[72] In the middle years, airline MRO providers tended to be most competitive because their substantial inventory and geographic presence gave them the ability to serve customers around the clock. Finally, when the system went out of production and out of most inventories, specialized independents were often the least-expensive suppliers of MRO services. Independents achieve economies of scale by purchasing bulk inventories from airlines or OEMs, obtaining licenses from OEMs to maintain and repair specific systems, and specializing in state-of-the-art inventory control to reduce costs.

---

[72]In addition, OEMs were—and still are—generally the only ones able to conduct major repairs because airline overhaul/maintenance (OH/M) facilities and independent vendors did not have the necessary equipment or training. For example, in the case of an inertial navigation system, only the OEM has the capability to calibrate the gyroscope. Simple electronic failures on the motherboard, on the other hand, can be fixed by the airline's OH/M facility or by a licensed vendor.

As the pressures of "must cost" continue to mount, however, this pattern of comparative advantage is beginning to change. Today, many large airlines no longer wish to accept standard warranty and maintenance clauses in new aircraft purchase contracts. Instead, they are asking for significant price reductions on new systems in return for reduced warranty coverage (in scope and in time), smaller inventories of spare parts, and fewer support services.[73] The result initially has been to put OEMs in a difficult position. With reduced warranty coverage, many of the costs associated with faulty systems are directly borne by the airlines. If an airline feels these costs are too high and that a system is unreliable, it might decide to switch suppliers. This creates an uncertain situation for the OEM, because it does not know what reliability threshold will be used to assess its performance. In response to this uncertainty, many OEMs felt obligated to offer services or replacement units not covered by warranty. In effect, OEMs found themselves offering airlines reduced prices on new equipment with no corresponding reduction in warranty coverage.

To get out of this unprofitable situation, many OEMs are replacing their price-plus-embedded-warranty contract structures with separate contracts for new equipment and for MRO services. Instead of treating MRO services as a cost of doing business, they now look at the after-sales market as a business opportunity.[74] They are linking their repair costs to the implicit prices of the warranties they provide to determine whether a warranty on a particular system is profitable. This makes good business sense for at least three reasons. First, the MRO business is larger than the business of aircraft production. Second, it is more profitable.[75] Third, for many OEMs it holds better

---

[73]One reason why the airlines are taking this approach is that the reliability of aircraft parts and equipment, as measured by Mean Time Between Unexpected Removals (MTBURs) and Mean Time Between Failures (MTBFs), continues to improve dramatically.

[74]As Harry Stonecipher, President of Boeing and former CEO of McDonnell Douglas, commented in 1997: "Clearly there is creative ferment in the MRO world. . . . The old way of looking at MRO was as a *cost of doing business*; the new way is as a *promising business in its own right.* That is to say, a business that appeals to customers, a business that offers opportunities for growth and profit." (Italics in the text.)

[75]From December 1992 to December 1997, the average stock price for an index of 15 suppliers rose 28.7 percent each year, higher than the average for Boeing, at 22.8 percent, or United Technologies, at 10.8 percent (United Technologies is the holding

growth prospects than sales of new equipment. Indeed, for the 1998–2002 period, the forecast average annual growth for new aircraft demand and for DoD R&D and procurement is just 8 and 3 percent, respectively, compared with projected growth in MRO activities of 40 percent by 2005.[76]

An OEM that has become prominent in this business is GE Aircraft Engines. Its entry into MRO started in 1992 with the acquisition of British Airways' engine overhaul and maintenance operations, quickly followed by a number of acquisitions and joint ventures. By 1998, GE's share of the engine overhaul market not serviced internally by airlines themselves had grown from 20 to 50 percent. Aftermarket activities now account for 40 percent of GE Aircraft Engine's revenues and 75 percent of its operating profit. Not surprisingly, such success is attracting other OEMs to the MRO market. Recently, Boeing and Airbus have indicated an interest in entering the MRO market as well (Schneider, 1998).

Within specified utilization parameters, many MRO service contracts now cover all activities related to a given aircraft, including entire aircraft systems "bumper-to-bumper." Sometimes, the airline is simply charged a per-hour utilization fee. Contracts based on per-hour utilization fees are particularly popular for aircraft engine maintenance, for which condition monitoring data are easily generated and wear-modes identified. In "power-by-the-hour"–type engine maintenance contracts, for example, aircraft operators pay a fixed monthly fee based on the hours flown within the month multiplied by a specific dollar rate per engine flight-hour.[77]

---

company for Pratt & Whitney). Firms included in the supplier stock index are BE Aerospace, BF Goodrich, Coltec, Ducommon, Curtiss Wright, Fairchild, HEICO, Hexcell, Moog, Precision Castparts, Sequa, SIFCO, Simula, Sundstrand, and Wyman Gordon. See Schneider (1998).

[76]New aircraft demand and DoD procurement numbers from Schneider (1998). MRO numbers based on a survey of MRO suppliers described in *AWST* (1997).

[77]See, for example, Rolls-Royce Allison (undated). "Power-by-the-hour" is also used to describe arrangements whereby airlines lease an engine from the OEM and the OEM is responsible for all maintenance. According to Schneider (1998), engine leasing is growing in popularity among airlines.

Airlines as well as OEMs perceive major advantages in these types of long-term fixed-price contracts for MRO services.[78] First, by establishing a fixed price for life-cycle MRO services, it removes any uncertainty about the future availability and price of spare parts and spare systems. Parts proliferation is not a problem because the OEM has responsibility for inventory control. Second, it creates an incentive for the OEM to improve the life-cycle reliability of the system. Improved system reliability reduces MRO service costs, which under fixed-price contracts translates directly into higher profits for the OEM.

In an effort to reduce their costs on MRO contracts, for example, OEMs have been tracking the number of so-called "No-Fault-Found" (NFF) removals—the removal of a component later found to be fully functional. Aviation data suggest that there are in excess of 400,000 such removals per year, representing 23 percent of all (1.76 million) component removals. With an average cost of $800 per removal—including labor, tracking, testing, etc.—NFF removals cost the industry approximately $300 million per year.[79] One of the best ways to limit such NFF removals is to staff line maintenance crews with expert technicians. These personnel are now often trained at or directly provided by OEMs.[80]

Even more important, throughout the life of a fixed-price MRO contract OEMs now have strong incentives to track actual system failures and weaknesses. Within the limits of FFF parameters, OEM system designers use information about items that fail repeatedly to identify areas for improvement. New technologies or designs are introduced to decrease production costs as well as increase reliability. Feedback from repair services provides a crucial element in the drive toward continuous product improvement.[81]

---

[78]MRO contract lengths and the range of contract features offered can vary considerably. More research is needed in this area.

[79]Approximately a third of this amount is related to avionics components.

[80]Industry studies reveal that NFF can be reduced by 68 percent using expert diagnostics, representing annual savings of $200 million (Canaan Group, 1998).

[81]The incentive to upgrade an item is constrained by the desire to avoid changes that require the costly process of FAA recertification.

To facilitate this process of continuous improvement, however, OEMs have had to undergo a major restructuring of their operations. At many firms, what were two entirely separate departments, R&D and support engineering, are now being integrated. Financial accounts are being restructured to allow OEMs to track R&D, production, and after-sale costs for individual systems. This kind of restructuring has become possible only since R&D risk-sharing has given OEMs increased control over the R&D agenda. The ability to control technical tradeoffs is helping to make the switch to "bumper-to-bumper" support both feasible and financially enticing for OEMs.

The competitive pressures introduced by airline deregulation have led to far-reaching changes in the market for commercial aircraft MRO services. More and more, airlines are choosing to outsource their MRO activities to third-party airlines, independents, and especially OEMs, including both system and airframe integrators. OEMs, for whom MRO services were once simply a "cost of doing business," are becoming major players in the MRO field. The trend toward OEM provision of "bumper-to-bumper" life-cycle service contracts has given them a new focus on product reliability. As a result, OEMs are now integrating their after-sale activities with R&D and production, introducing new designs and technologies that hold the promise of increased cost-effectiveness.

For DoD, these after-sale market trends in the commercial world also hold the promise of increased weapon system reliability at lower cost. Outsourcing life-cycle support services to OEMs for a fixed price could help to ensure system supportability and maintainability over time. Although there are some real concerns about the support and maintenance of military aircraft (and weapon systems more generally) by civilians during wartime, experience suggests that these obstacles to increased reliance on commercial suppliers are not insurmountable.[82] The gains could be large—not only in terms of lower life-cycle support costs but as an integral part of the CMI strat-

---

[82]According to Stonecipher (1997), during the Gulf War McDonnell Douglas had 1100 technicians supporting F-15s in Saudi Arabia. When the C-17 was deployed to Bosnia, support personnel from McDonnell Douglas were again deployed.

egy of increasing contractor configuration management and control of weapon system programs.[83]

CONCLUSION

We began this chapter by asking four questions about the full or partial adoption of a commercial-like approach to weapon system acquisition:

1. Can system cost be reduced without sacrificing performance?

2. Will qualified contractors be willing to absorb more of the market and technical risk associated with new aircraft development?

3. Can DoD promote and maintain adequate levels of price-based competition?

4. Can DoD ensure the supportability and maintainability of systems over time?

Based on an examination of the ways in which U.S. participants in the market for large transport aircraft have approached similar questions, our answer to each is a qualified "yes." On net, we believe that adopting a commercial-like acquisition strategy will prove beneficial to DoD. If military contractors adopt the "best" commercial practices used by their commercial counterparts, we expect to see a decline in the cost to DoD of developing, producing, and maintaining military aircraft. We caution, however, that cost declines may be accompanied by a diminution of the technical virtuosity of U.S. military aircraft if too much emphasis is placed on cost-control relative to performance requirements.

We found that binding cost constraints introduced by "must cost" have shifted the focus of airlines and aircraft manufacturers from performance to cost. This has not resulted in airliners with poor performance characteristics; in fact, along certain dimensions there have been notable improvements. Arguably, however, aircraft

---

[83]There are serious and possibly unique difficulties involved with designing appropriate warranties for weapon systems, however, as discussed in Kuenne et al. (1988). Relevant RAND studies on how best to source and structure military logistics support services include Keating (1996) and Keating et al. (1996).

manufacturers have been less willing to introduce dramatic technological improvements. Instead, the focus of technical innovations
has been to meet the joint goals of low operating costs and super-
safe flight in increasingly crowded skies. If DoD adopts a true "must
cost" approach, emphasizing cost over other considerations, then
careful program management will be required to maintain technical
innovation in the desired areas.

We also expect that DoD will be able to lower its financial support for
new aircraft development without causing qualified contractors to
leave the market. Contractors will respond to the challenge by
choosing partners who are able to provide their own financial capital
as well as technical capabilities. To facilitate this, DoD must be willing to give prime contractors and their risk-sharing partners increased responsibility for and flexibility in program management,
working closely and cooperatively with them as part of IPTs. DoD
may also be required to accept a narrowing of its supplier base, as
prime integrators choose from among a smaller group of qualified
suppliers with proven track records.

To maintain sufficient competition to prevent price-gouging, as well
as to encourage greater risk-sharing by contractors, DoD should take
three steps. First, DoD should continue to foster cooperation between the services in weapon system acquisition, not only to take
advantage of sharply declining average unit production costs but
also to encourage manufacturers to compete for military contracts.
Second, with respect to standardization and interoperability, DoD
should encourage the use of existing commercial specifications and
standards wherever practical, but take advantage of growing commercial competency in modular integrated systems when the associated economies of scope and scale are large. For these types of
systems, the pros and cons of life-cycle "bumper-to-bumper" support service contracts provided by OEMs are worth examining more
carefully. Third, DoD must carefully consider the role of foreign
firms in the development of sensitive technologies. In the commercial world, foreign firms not only provide important financial and
technical resources to their U.S. partners but also provide the world's
airlines with a potent alternative to the dominance of U.S. firms.
Competition from abroad deters U.S. manufacturers from resting on
their laurels, encouraging them to find new ways to cut costs and
pass the savings on to their airline customers. As a final considera-

tion, military producers will know that if they exorbitantly exploit any changes in cost-control regulation, political pressure will certainly be brought to bear to reinstate the controls.

Chapter Seven

PILOT PROGRAMS: LESSONS LEARNED

INTRODUCTION

In Chapter Six, we examined commercial markets to glean insights into improved commercial-like approaches to military acquisition. We began by noting that key differences between commercial and government military markets make comparisons across these markets problematic. However, certain commercial markets, such as the commercial transport aircraft and aircraft equipment markets, exhibit many characteristics similar to the government military aerospace market. We showed how, particularly after deregulation of the airline industry in the late 1970s, the commercial aircraft market adopted a strategy of "must cost," in conjunction with mechanisms such as contractor risk-sharing and IPTs, to cope with potential performance shortfalls and price gouging in an increasingly price-sensitive environment. We concluded by suggesting that many of these "best" business practices may be exploitable in the military market.

Yet it can still be argued that there remain important differences between the commercial aircraft and the military aircraft markets—most notably the existence (in most cases) of a single buyer in military markets and the tendency (or necessity) of military programs to incorporate high-risk technologies to achieve the highest possible performance. These differences continue to raise serious questions in the minds of some CMI opponents as to the applicability of commercial market mechanisms to military acquisition programs.

The government has responded to these concerns by testing commercial-like approaches to military acquisition in a variety of innovative pilot and demonstration programs. And although some of these programs have been examined extensively in isolation, we are not aware of any studies that attempt to determine if they provide any across-the-board "lessons learned." This chapter takes such an approach, attempting to assess to what extent, and with what success, commercial-like approaches based on market mechanisms are being applied to military programs and what can be learned from them.

Two categories of programs, representing systems from munitions to potential weapon system platforms, were selected for review. The programs entail full-scale R&D and production as well as modification of existing systems. The first category, which is the only one reported on in detail here, includes three of the most important acquisition pilot programs currently under way. These programs aim at the development and production of a new generation of "smart" munitions for the U.S. Air Force and Navy: The Joint Direct Attack Munition (JDAM), the Wind Corrected Munitions Dispenser (WCMD), and the Joint Air-to-Surface Stand-Off Missile (JASSM). These official acquisition reform pilot programs have progressed beyond the concept development stage and (1) focus on developing military-unique combat systems from their inception under the direction of the user services; (2) were from the beginning intended to result in the full development, procurement, and operational deployment of actual weapon systems; and (3) make use of virtually every acquisition reform measure and concept proposed over the last several years.

The second category is made up of two programs initiated by the Defense Advanced Research Agency (DARPA) for the development and possible production of high-altitude endurance (HAE) unpiloted air vehicles (UAVs), plus an innovative modification program, DoD's Commercial Operations & Support Savings Initiative (COSSI). The UAV programs are the Tier II+ Global Hawk under development by Teledyne Ryan Aerospace and the stealthy Tier III– DarkStar under development by Lockheed Martin and Boeing.[1] Both of these pro-

---

[1]The DoD cancelled the DarkStar program in early 1999.

grams have been designated as Advanced Concept Technology Demonstration (ACTD) programs and are operated under DARPA's Section 845 Other Transactions Authority.[2] The COSSI program is a joint Army-Navy-Air Force program that encourages the insertion of commercial technologies into military systems to lower long-term operations and support costs without degrading system performance. Like the DARPA programs, COSSI also operates under the Other Transactions Authority for prototypes.

The two DARPA programs contain many novel characteristics similar to formal Defense Acquisition Pilot Programs (DAPPs).[3] Nevertheless we comment only briefly on them at the end of this chapter, in part because they are technology demonstration programs not administered by the services and not necessarily intended to lead directly to procurement of operational systems, and in part because they are reported on in other RAND research.[4]

MUNITIONS PILOT PROGRAMS[5]

JDAM

JDAM is an early trial program for testing key aspects of the Clinton Administration's defense acquisition reform measures. Indeed, Lt Gen George Muellner, former Principal Deputy Assistant Secretary of the Air Force for Acquisition, characterized JDAM as "the linchpin" of "the broader Department of Defense's acquisition streamlining

---

[2]In principle, ACTDs are intended to allow DARPA, in close association with potential user services, to rapidly integrate relatively mature technologies into prototypes to demonstrate a useful operational capability. Section 845 Other Transactions Authority eliminates nearly all normal procurement statutes and FARs to permit maximum program flexibility in developing demonstration prototypes of weapon systems. See DoD (1998).

[3]For example, both UAV programs have in principle a hard "must cost" unit fly-away price of $10 million in FY94 dollars, while all other aspects of the program are flexible and can be traded off against cost. The unit fly-away price is defined as the average price of air vehicles 11–20, including sensor payload, for both programs.

[4]See, for, example Drezner and Sommer (1999) and Sommer et al. (1997). After completion of the DARPA-run technology demonstration phase, these programs are to be handed off to the Air Force for full-scale development.

[5]Most of the information on these three programs was acquired from open sources, program documents, and interviews with the Program Offices (at Eglin Air Force Base, Florida), and with contractors.

activities" (Muellner, 1996). JDAM is an Acquisition Category (ACAT) 1D program, the most important Air Force acquisition category.[6]

JDAM originated as a traditional military acquisition program. Nonetheless, from the very beginning, the Air Force imposed a high-priority average unit price target of $40,000. In 1994, DoD designated JDAM as an official DAPP under the 1994 Federal Acquisition Streamlining Act (FASA), which mandated a wide variety of acquisition reform measures.[7] Dr. Paul Kaminski, sworn in as Under Secretary of Defense for Acquisition and Technology in October 1994, supported JDAM as a major test case for acquisition reform.

The JDAM program aims at developing sophisticated—but affordable—"strap-on" guidance kits that can be attached to standard Mk-83 and BLU-110 1000-lb "dumb" bombs, and Mk-84 and BLU-109 2000-lb "dumb" bombs. Through the use of an inertial navigation system augmented by updates provided by GPS, which guide active control surfaces, JDAM kits permit highly accurate delivery of bombs from a variety of aircraft platforms under a wide range of adverse weather and environmental conditions.[8] JDAM has a range of about 15 nautical miles when dropped from high altitudes.

---

[6]ACAT 1D programs are Major Defense Acquisition Programs (MDAPs). According to the Defense Systems Management College, "An MDAP is defined as a program estimated by OUSD/A&T to require eventual expenditure for research, development, test, and evaluation of more than $355 million [fiscal year (FY)96 constant dollars] or procurement of more than $2.135 billion (FY96 constant dollars), or those designated by OUSD/A&T to be ACAT I." ACAT 1D programs are those where the Milestone Decision Authority resides at the highest level possible: OUSD/A&T.

[7]The DoD Authorization Act for FY94 designated five programs as statutory DAPPs: JDAM, Fire Support Combined Arms Tactical Trainer, Joint Primary Aircraft Training System, Commercial Derivative Engine, and Nondevelopment Airlift Aircraft (later dropped). FASA provided regulatory relief for these programs and gave authorization to treat them as commercial procurements. Later, the C-130J and the Defense Personnel Support Center were added as "regulatory" DAPPs. See OUSD/AR (1997a) and OUSD/A&T (June 30, 1997).

[8]U.S. and allied forces used a wide variety of existing "smart" munitions during Desert Storm combat operations in Kuwait and Iraq, often with great effect. However, many current smart munitions guidance kits use electro-optical, laser, or infrared sensors whose performance can be degraded in poor weather conditions, when the battlefield is obscured by smoke and dust, or by other factors. The requirement for JDAM and WCMD arose from the need to develop munitions guidance kits for unguided munitions that could operate well in all weather conditions and in other situations where visibility is poor.

The JDAM configuration and baseline weapons are shown in Figure 7.1.

WCMD

The Air Force WCMD program has some similarities to the JDAM effort. In response to FASA and DoD's acquisition reform, the Air Force designated WCMD a "lead program" to test out acquisition reform within the Air Force. WCMD is the only Air Force acquisition reform "lead program" for a totally military-unique combat weapon system developed from scratch.[9]

Compared to JDAM, WCMD is a somewhat simpler tail guidance retrofit kit employing an inertial navigation unit and active control

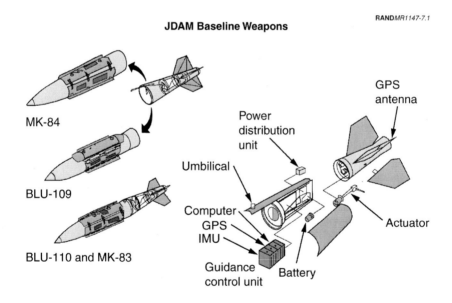

Figure 7.1—JDAM Baseline Weapon System

---

[9]WCMD is one of four Air Force "lead programs" selected to implement acquisition streamlining initiatives. The other three are Evolved Expendable Launch Vehicle, Ground Theater Air Communications System, and Space-Based Infrared Systems.

surfaces intended for use on three "dumb" air-dropped munitions dispensers: the CBU-87/B Combined Effects Munition, the CBU-89/B Gator, and the CBU-97/B Sensor Fused Weapon. WCMD kits are intended to enhance aircraft survivability by permitting a GPS-capable aircraft to drop munitions dispensers from medium altitudes with accuracies equal to or better than those currently achieved through dangerous low-level attack profiles. WCMD's inertial measurement unit (IMU), which can be updated with GPS-quality data from the launch aircraft, corrects for launch transients and wind deflections, thus providing medium-altitude all-weather capability. Its active control surfaces and wind estimation and correction software help WCMD achieve a target accuracy of 85-ft CEP (Circular Error Probable) from altitudes up to 45,000 ft.

JASSM

JASSM is the largest and most sophisticated of the three programs. Like JDAM, JASSM is a joint Air Force–Navy project with the Air Force in the lead role. However, JASSM is a much more complex autonomous stand-off munition. It is a long-range powered cruise-missile with stealthy characteristics. Like JDAM, the missile is equipped with an inertial navigation system and a GPS receiver for navigation. In addition, however, JASSM adds a sophisticated autonomous terminal guidance and automatic target recognition system for true stand-off fire-and-forget capability. JASSM will have a range in the hundreds of miles depending on the launch platform and altitude. With overall performance objectives similar to the ill-fated Tri-Service Standoff Attack Munition (TSSAM) (described below), JASSM is a technologically challenging program, particularly in overall system integration, autonomous guidance, and automatic target recognition.[10]

DoD approved development of JASSM in September 1995, designating it a "Flagship Pilot Program" for acquisition reform.[11] Former Assistant Secretary of the Air Force for Acquisition Arthur L. Money characterized the JASSM program as employing "an aggressive ac-

---

[10]See the description in GAO (June 1996).

[11]More accurately, JASSM is a "Flagship Pilot Program for CAIV."

quisition approach using virtually every acquisition reform initiative known to date." (Quoted by Chapman, 1996.) As an ACAT 1D program, JASSM, like JDAM, is also in the highest Air Force acquisition category.

MEASURES EMPLOYED TO ACHIEVE CMI GOALS

As noted in the introduction to this report, the primary objectives of a more commercial-like approach to military acquisition, or CMI, are to:

- Reduce costs of acquiring and supporting weapon systems

- Improve performance at Initial Operational Capability (IOC) and throughout the life-cycle of a weapon system

- Shorten development times

- Improve reliability and maintainability

- Help maintain the defense-relevant portion of the industrial base.

The three Air Force munitions programs examined here, as well as other DoD acquisition reform pilot programs, employ a variety of reform measures intended to ensure the achievement of these goals. Many of the measures are drawn from or attempt to replicate conditions in commercial markets. For example, they promote the use of commercial parts and technologies, and encourage the participation of commercial firms, in order to reduce costs and raise quality. At the same time, they are designed to incorporate the type of market-driven safeguards that act to ensure fair pricing and high quality in most commercial markets. These reform measures can be grouped under four general headings:

- Reduction of the regulatory and oversight burden

- Requirements reform and implementation of "must cost" concepts such as CAIV

- A more "commercial-like" R&D program structure, incorporating elements of both "textbook" and "best" commercial practices

- A more "commercial-like" purchase and support of developed systems based on performance incentives.

Reduced Regulatory and Oversight Burden

As discussed in Chapter Two, a key component of CMI is the reduction of the government-imposed regulatory and oversight burden that results in higher costs for DoD—many argue with little value added—and that discourages commercial firms from doing business with the Defense Department. A central purpose of the DAPPs established by the 1994 FASA legislation mentioned above, as well as other pilot programs, was to test outcomes of programs that had been granted relief from the regulatory burdens typically imposed on contractors by the government. In essence, DAPPs permitted DoD to experiment with purchasing a military system as if it were a commercial product. FASA granted DAPPs statutory exemptions that it had mandated for DoD purchases of pure commercial items and permitted DAPP program offices to seek waivers of other statutory regulations from the Secretary of Defense. Thus JDAM, WCMD, and JASSM all enjoy significant regulatory and statutory relief.

JDAM began as a traditional ACAT 1 program. After its designation as a DAPP, DoD eventually granted 25 FAR and 25 DFARS waivers (OUSD/AR, April 1997; SAF/AQ, 1997). These waivers permitted the program to be managed more like a commercial business relationship between buyer and seller/developer. Although cost and other data reporting were still required, the government accepted high-level data in contractor format for most data submittals, program reviews, design reviews, and accounting audits. Formal program and contractor oversight requirements were reduced, and the government greatly reduced the complexity of contract requirements and the Statement of Work (SOW). Contract Data Requirements Lists (CDRLs) were reduced from 250 pre-DAPP JDAM RFPs to 28.[12] The program office agreed to accept CDRLs in contractor format. In

---

[12]By the accounting of the Under Secretary of the Air Force for Acquisition (SAF/AQ), the number of CDRLs was reduced from 146 to 22 if the comparison is made between the RFP release in August 1993 (FY95 President's Budget) and December 1997 (FY99 President's Budget). The remaining CDRLs were required mainly by the test and safety communities.

addition, the number of pages in the RFP fell from 986 to 285. The contractor SOW was changed to a Statement of Objectives (SOO) and reduced from 137 pages to 2 pages.

DoD experts estimated that as a result of regulatory relief, government contract administration hours required by the Defense Contract Management Command (DCMC) and its onsite representatives on a three-year R&D contract for JDAM declined by more than three quarters from over 20,000 to under 5000 hours. The JDAM Joint Program Office estimated that because of regulatory relief the total contract administration hours required for the program at the end of 1997 had decreased by nearly 40 percent.

Similar results are being reported for WCMD and JASSM. The WCMD RFP contained only 93 pages. The WCMD contractors developed and wrote their own SOWs, system specification, data requirements documents, and integrated master schedules. Only 18 CDRLs are on contract for WCMD. Although the Air Force uses C/SCS on WCMD, contractor costs during R&D are reported largely in contractor format and down only to the subsystem level, not the piece part level. In the case of JASSM, the government reduced the RFP even further—especially considering the greater complexity of the system—to only 78 pages. Government and industry cooperated closely to develop the original draft RFP on both WCMD and JASSM. Formal contractor proposals were limited to 320 pages on JASSM, including the system performance specification and integrated master plan and schedule. Cost data were limited to 10 pages. One-page SOOs replaced huge SOWs. The first-phase R&D contract for JASSM contained only 16 CDRLs. However, the government did impose CAS on the prime contractors.

It is difficult to quantify the cost savings enjoyed by either the contractors or the government as a result of reductions in the oversight and regulatory burden. Indeed, some studies suggest that the direct cost savings that accrue to contractors from reduction of reporting requirements and other regulatory compliance burdens are modest. However, an additional strong justification for reducing the regulatory and oversight burden is to make defense contracts more attractive to commercial firms and to encourage insertion of commercial parts and components. For this reason as well as cost savings, meaningful efforts were made to reduce the regulatory burden in all three of the munitions programs studied here.

For example, the government ended up imposing CAS on the JASSM prime contractors. But the SPO worked closely with the Office of the Secretary of the Air Force to overcome objections by potential commercial vendors and subcontractors to the imposition of CAS and to various RFP flowdown clauses. As a result, vendors and subcontractors who signed FFP contracts were exempted from CAS. Using the revised rules on contracting for commercial items, the SPO was able to reduce the number of flowdown clauses to the appropriate subcontractors to only four.

These reductions in regulatory and reporting burdens appear to have contributed to the ability of the prime contractors to subcontract more often to commercial vendors and to use much less expensive commercial parts and technology, thus apparently reducing the production costs to the government of their developed systems. This issue is discussed in greater detail in some of the following sections.

Requirements Reform

JDAM, WCMD, and JASSM are pioneering attempts at promoting CMI through requirements reform, including full implementation of CAIV processes. Each seeks to reduce costs by, first, using carefully crafted mission requirements that avoid gold plating and unnecessary capabilities; second, presenting system requirements to contractors in terms of mission performance rather than detailed design and technical specifications; third, minimal use of Mil-Specs; and fourth, contractor configuration control during R&D.

In the official Operational Requirements Documents given to contractors, all three programs replaced detailed technical specifications and "how-to" design-and-build directives with broad mission performance objectives. A few of these objectives were identified as nonnegotiable and were stated in terms of broadly defined "Key Performance Parameters" (KPPs). They were often defined in terms of multiple performance measures—which might be tradeable among themselves—but most had certain minimum "threshold" requirements that had to be met. The remaining objectives were identified as "nice-to-have" but not "must-have." The purpose of this categorization was to prioritize objectives to focus contractor efforts on the most important program requirements and, where appropriate, to facilitate and encourage tradeoffs for cost and other reasons.

Originally, the program planners intended to require no Mil-Specs whatsoever so that contractors could exploit off-the-shelf commercial technologies and parts to reduce costs. However, a few Mil-Specs were eventually adopted to ensure compatibility with host aircraft and for safety considerations. For example, the weapon stores and software interface with the host aircraft required the use of Military Standard (Mil-Std) 1760, whereas communications between the JDAM onboard processor and the host aircraft necessitated use of the Mil-Std 1553 high-speed bus. JASSM and WCMD experienced similar additions of some Mil-Specs.

Nonetheless, these programs show a dramatic reduction in Mil-Spec requirements compared with traditional programs. In the case of JDAM, the baseline pre-DAPP RFP included 87 Mil-Specs, compared with just a handful in the DAPP phase. Interestingly, JDAM also did not require any specific commercial specifications or standards, nor were any Mil-Specs or commercial standards imbedded in a SOW, because only a SOO was required from the contractors. WCMD eliminated all but two Mil-Specs.

All three programs remained close to the original intent of using only broad mission performance requirements instead of detailed technical specifications. In the case of JASSM, the Air Force and Navy user communities agreed that only three objectives were nonnegotiable KPPs: range, missile effectiveness, and aircraft carrier compatibility. Measures of merit for missile effectiveness were carefully developed and clearly communicated to contractors.[13] In addition, the government developed seven performance objectives that were considered desirable, but that contractors could trade off against each other and against other factors to reduce costs.

---

[13]The government gave the contractors the official JASSM Operational Requirements Document. The government used a computer simulation to measure missile effectiveness. All contractors had access to the model and could use it to test their technical proposals. Furthermore, contractors could test and question the methodology, tools, and assumptions built into the model.

Yet, the JASSM contractors were not told *how* to achieve KPPs such as "missile effectiveness" while achieving the $400,000 production price target. For example, one defined characteristic of the "missile effectiveness" parameter was survivability, which could be achieved by increasing missile speed, lowering radar cross-section, or by a variety of other means. Or survivability could be traded off against other defined characteristics such as reliability or probability of damaging various types of targets, or against cost. It was up to the contractor's engineers to use creative new approaches to try to optimize the tradeoffs between a whole variety of factors, meet the target price goals, and convince the customer that the correct design decisions had been made.

KPPs did often contain some "threshold" requirements that were nonnegotiable. An example of a threshold requirement was missile compatibility with the B-52, F-16, and F/A-18E/F.[14] JASSM missile compatibility with a variety of other aircraft was labeled as an objective. Whereas all contractors had to achieve missile compatibility with the F-16, each contractor reached its own conclusions about the cost benefits of not achieving compatibility with, for example, the F-117. This helped the government and the contractors to understand clearly the cost of nonessential operational requirements and to decide if that capability was really worth the cost. The iterative process between contractors and the government led to changes in emphasis and priorities in the overall system requirements.

A fundamental JASSM program requirement that could not be traded, and that is enshrined in the Operational Requirements Document, is the "must cost" price ceiling of $700,000 in FY95 dollars for the Average Unit Production Price (AUPP) of JASSM, and the "should cost" price objective of $400,000 (FY95$). This requirement emerged in 1995 after the cancellation of the TSSAM program. Begun in 1986 by Northrop, TSSAM aimed at providing a stealthy, long-range cruise missile with autonomous terminal guidance and target recognition capabilities. However, after many years of development, the Pentagon cancelled TSSAM because the program was plagued with reliability problems and high costs. At the time of pro-

---

[14]The F/A-18E/F was later dropped as a threshold aircraft.

gram launch, government officials estimated an average unit production cost for TSSAM of $728,000 in then-year dollars.[15] By 1994, average unit production costs for approximately 2500 missiles were expected to exceed $2 million in then-year dollars. The Office of the Secretary of Defense (OSD) concluded that the program had to be cancelled because of excessive cost, and that a cost above $700,000 (FY95$) for a TSSAM follow-on missile would prevent procurement of adequate numbers of missiles. The cancellation of TSSAM and the continuing critical need for an affordable long-range low-observable stand-off missile with performance capabilities similar to TSSAM are the key reasons that OSD designated JASSM a flagship program for CAIV.[16]

The government structured the JDAM and the WCMD design phases in a similar manner, making sure that the contractors were presented with clearly prioritized requirements stated in terms of broad mission performance and emphasizing low cost as a key requirement. For example, the government formulated the following seven KPPs for JDAM:

- Target impact accuracy of 13 meters CEP with GPS

- Accuracy unaffected by weather conditions

- In-flight retargeting capability (before release)

- Warhead compatibility[17]

- Carrier suitability

---

[15]Under cost-plus–type production contracts, DoD's relevant measure of cost is the average cost of production per unit plus the (positive) contractor profit allowed under DoD profit policy. With the switch to fixed-price contracts, "profits" may be either positive or negative, and the relevant measure is the average price DoD pays per unit of production.

[16]A requirement that drove up costs on TSSAM—and that was dropped on JASSM— was tri-service deployment capability. TSSAM had to be capable of launch from both Air Force and Navy aircraft, as well as from Army ground launchers. This requirement raised numerous technical difficulties for the TSSAM developers. JASSM dropped the Army ground-launch requirement and retained only the Air Force and Navy air-launch requirement.

[17]JDAM guidance kits had to work with the Mk-84 general-purpose 2000-lb bombs, the BLU-109 2000-lb penetrating bomb, and the BLU-110/Mk-83 general-purpose 1000-lb bombs.

- Primary aircraft compatibility[18]
- AUPP of $40,000 or less.[19]

The seventh KPP, an AUPP of $40,000 or less, was based on calculations that a JDAM-type weapon under a traditional acquisition approach would have an AUPP for 40,000 units of $68,000.[20] Because of budget limitations, senior Air Force officials concluded that JDAM could not be procured in adequate numbers at this price. An AUPP of $40,000 was the maximum that the government would be able to pay.

The focus on cost, the use of broad mission requirements, the emphasis on cost/benefits tradeoffs, the lack of Mil-Spec requirements, and the control of the contractor over configuration and technical solutions for the JDAM program produced some dramatic results. Several contractors took the initiative to exploit commercial technologies, insert COTS parts and components, develop creative technical solutions, and trade off performance against cost where appropriate to achieve significant cost reductions.

An interesting example is that of a JDAM power transistor proposed by one of the competing contractors for use with the braking mechanism on the shaft of the torque motor that turns the fins that steer the bomb. The original requirement called for a 2500-inch-pound stall torque capability. This necessitated the use of an expensive Mil-Spec power transistor that cost $25 per part. But because of the strong emphasis on cost reduction, and the freedom granted the contractor to try different and innovative non–Mil-Spec solutions, one contractor proposed redesigning the unit in a way that it argued maintained overall performance but lowered the stall torque requirement to 1600 inch-pounds. This meant a less-capable Mil-Spec

---

[18]Four aircraft (F-22, B-52H, F/A-18C/D, and AV-8B) were listed as threshold requirement aircraft (1000-lb bomb versions only for F-22 and AV-8B), which means this capability requirement had to be met. Compatibility with nine other aircraft (B-1, B-2, F-16C/D, F-15E, F-117, F/A-18E/F, F-14A/B/D, P-3, and S-3) was listed as an objective. Compatibility with the objective aircraft was a requirement fully subject to tradeoff analysis with cost and other factors.

[19]For 40,000 units in FY91 dollars; in FY93 dollars, the AUPP was $42,175 (adjusted using the GDP deflator).

[20]In 1993 dollars.

power transistor could be used that cost only $15. The contractor then decided to examine commercial power transistors. Eventually a COTS part was proposed that cost $4.05.

By redesigning the component and qualifying a COTS part, the contractor argued that more than $500 per guidance unit could be saved, because each unit contains 24 power transistors. With a planned buy of 87,495 units, this could have resulted in a potential production cost savings of almost $44 million.[21]

Another example provided by the JDAM SPO is the shipping and storage container for the JDAM guidance kit. This container must be extremely rugged to guarantee full operational capability of the guidance kit after years of storage and transport in extreme environmental conditions. Traditional Mil-Spec containers of this type have one guidance kit per container, are made out of aluminum, and cost about $1600. Originally JDAM had separate Navy and Air Force requirements for the container. The Navy requirement was more demanding because of harsher conditions for shipboard storage and use. The Navy requirement in September 1994 called for a two-kit aluminum container that would cost an estimated $2200. The Air Force requirement also called for a two-kit container.

Trade studies were conducted between September 1994 and February 1995 on both container designs. To save costs, the Air Force agreed to substitute high-density polyethylene (HDP) plastic for aluminum, lowering the cost to $800. Later, the Navy agreed to accept a front-opening fiberglass container costing $1100. Meanwhile, the Air Force had moved toward a commercial electronics storage philosophy and was willing to accept a much simpler HDP plastic design using a vapor barrier bag. This container cost $600. Finally, in August 1995, McDonnell Douglas (now Boeing) came up with a design acceptable to both services that was made out of fiberglass and used a vapor barrier bag. McDonnell Douglas essentially derived this technology from the shipping and storage technologies found in the commercial electronics industry. The final container design had an estimated cost of $600, or $300 per guidance

---

[21]This design was proposed by one of the losing contractors, so this COTS power transistor was never actually fully tested or incorporated into the final JDAM design for the fin braking mechanism.

kit, compared with the traditional Mil-Spec container at $1600 per guidance kit.

A final example is particularly interesting because one of the contractors directly challenged the cost-effectiveness of a specific system performance requirement and won his point. The contractor pointed out that by slightly reducing the required low-altitude captive carriage time in the worse-case scenario in which JDAM was intended to operate, a design change could be made that significantly reduced costs. The contractor argued successfully by using the results of the government combat model, which showed a relatively small decrease in overall combat effectiveness for a large decrease in design complexity and cost.

According to Boeing St. Louis, more than 200 cases of detailed trade-off studies that reduced JDAM costs have been formally documented in the program's Affordability Trade Studies, although most of the specific cases are proprietary. One of these cases is discussed below as an example of insertion of dual-use technology.

Because of the emphasis on cost promoted by CAIV, the tradeoff analysis of performance versus cost that CAIV encouraged, and the elimination of most Mil-Specs, JDAM is able to make extensive use of COTS processors, boards, chips, and other commercial parts and components. Originally program officials and contractors had planned to acquire major subsystems and components from commercial sources or production lines. Table 7.1 lists the planned sources for various key components for the designs of the two competing contractors during the final competition phase. In the case of the Boeing design, the IMU, the GPS receiver, the mission computer (MC), and the control actuators make up 85 percent of the cost of the guidance kit. Although these subsystems are now acquired from military production lines, all of them contain commercial parts, are slightly modified versions of commercial items, are government off-the-shelf items (GOTS), or could be sold as commercial items.

For example, the Boeing JDAM mission computer, as shown in Table 7.1, was originally intended to come from a commercial source. Eventually, however, Boeing designed its own mission computer and selected Unisys (now Lockheed Martin Tactical Defense Systems) to manufacture the mission computer on a military production line.

Table 7.1

Commercial/Military Mix of JDAM Contractor Production Lines[a]

Item	Boeing[b]	Lockheed Martin
Integration/assembly	Commercial	Military
IMU	Military	Military
GPS receiver	Military	Commercial
Mission computer	Commercial	N/A
Circuit cards	Commercial	N/A
Connector	Commercial	N/A
Actuators	Commercial	Military
Power supply/distributor	Military	Commercial
Thermal	Military	Military
Container	Commercial	Military/Commercial
Fin	Commercial	Commercial
Tail	Military	Military/Commercial
Hardback/nose	Commercial	Military/Commercial

[a]As of late 1996. Boeing, the winning contractor, later switched the mission computer to a military production-line source. Sources for other items may also have changed.

[b]Formerly McDonnell Douglas, the winner of the Phase II contract.

Boeing's dedicated military mission computer is programmed using the Ada language, which is uncommon in the commercial world. Nonetheless, the mission computer's architecture is similar to computers that sit on many people's desks. At its heart is a Motorola microprocessor similar to the one that, prior to the JDAM program, was used by Apple Computer, Inc., as the basis for its Performa 470 series of personal computers. Boeing hopes to upgrade this chip with one similar to that used in the PowerPC or iMac.

Both JDAM and WCMD use the Honeywell HG1700 IMU, a highly miniaturized dedicated military IMU developed by Honeywell Military Avionics explicitly for applications such as smart munitions, UAVs, and missiles. Similar IMUs are used in commercial applications such as railroad vehicle control and landslide detection because of their low-cost and high-performance characteristics. Boeing and Honeywell implemented 11 design changes, or "affordability initiatives," to reduce production costs on what was already an inexpensive OTS item from the JDAM and WCMD perspective by a further 20 percent. These cost reductions were passed on (in part) to DoD.[22]

Boeing's Affordability Trade Studies document the reduction in cost of the Honeywell HG1700 IMU through commercial parts insertion. On the JDAM program, a Boeing/Honeywell IPT worked hard to reduce the cost of this item through the identification of cheaper commercial parts for insertion into the IMU, as well as through other reform initiatives. For example, the original HG1700 IMU connectors were expensive Mil-Spec parts. Eventually a way was found to use much less expensive Honeywell commercial IMU connectors, which saved about $100 per JDAM IMU. This change alone has the potential of saving millions of dollars in production costs over the planned JDAM production run.

The GPS guidance unit for JDAM is provided by Rockwell Collins Avionics and Communications Division, which is part of the Rockwell International's Defense and Electronics group. The GPS receive module is a deliberate variation on the company's widely used GEM III GPS embedded-module receiver, which is based on the Standard Electronic Module, Format E (SEM-E) standard. The redesigned receive module allows for greater spreading of the microprocessors, which in turn allows Rockwell to use less-expensive electronic parts.[23]

---

[22]Some of the initiatives included changes in make/buy decisions, parts changes, investment in cost-saving capital equipment, and using commercial inspection processes. One such initiative is discussed in greater detail below.

[23]SEM-E may help standardize military avionics modules, but it also may make it more difficult to adapt commercial modules based on commercial standards. Avionics modules used in munitions, missiles, and tactical fighter aircraft are constrained by special factors such as size, weight, high-vibration, and shock requirements. SEM-E is a form factor standard for electronic modules and connectors that is increasingly popular in military avionics. It is based on conduction-cooled technology, a small form factor, and a blade and fork style connector that provides high reliability in high-shock and vibration environments. It is not used widely in the commercial world, but it is compatible with open architectures. Using the SEM-E standard permits upgrades by ensuring that modules are compatible in size and connectors. However, according to one source, the SEM-E standard size (approximately 6 x 6 inches) may make military avionics modules increasingly incompatible with the commercial market's larger module sizes. Thus it may be more difficult to adapt commercial modules for airborne avionics use. See Defense Advanced Research Projects Agency (DARPA) (1995).

Both prime and subcontractors conducted extensive testing of non–Mil-Spec commercial and plastic-encapsulated parts and their applicability to the environmental conditions in which JDAM would operate. A temperature range of –55°C to +85°C was eventually accepted as the baseline standard for electronic parts. On the high end, this standard permits use of catalog COTS industrial- or automotive-grade parts. However, the low end surpasses the requirements for commercial parts and is indeed the same as the Mil-Spec standard. Therefore, commercial catalogue parts usually had to be tested and/or screened.

According to one Boeing JDAM official, the contractor's experience with testing commercial parts for insertion into JDAM subsystems was highly variable. Some suppliers conducted their own testing for Boeing at a relatively low cost. Other suppliers were willing to conduct tests at their own facilities but charged Boeing a substantial premium. A third category of suppliers agreed to sell testing devices or data to Boeing so that the prime contractor could conduct its own testing. Again, depending on the part or subcomponent, Boeing's cost of testing the commercial parts itself varied considerably. Finally, some suppliers agreed to sell commercial parts but refused to conduct the necessary additional testing required and would not provide the data or devices necessary for the prime contractor to conduct the tests.

Boeing officials claim that the extensive trade studies and commercial parts testing conducted during the initial phase of the program to identify appropriate commercial parts for insertion into JDAM proved to be an expensive and time-consuming effort. Nonetheless, the extra effort to qualify commercial parts seems to have paid off in much lower production costs. According to one account, the use of plastic encapsulated parts saved $535 per unit.[24] This is about 3 percent or less of the AUPP of the JDAM in 1998.

Similar incentives instilled by the CAIV approach, in combination with the virtual elimination of the need to use Mil-Spec parts and processes, produced similar results on the JASSM program: extremely creative and innovative approaches to exploiting existing

---

[24]Assistant Secretary of the Air Force (Acquisition), *JDAM—The Value of Acquisition Streamlining*, no date.

commercial and military technologies and parts to lower costs while still producing acceptable performance capabilities in a military environment. Two interesting examples are the process technologies chosen to manufacture the fuselage, the wings, and the vertical stabilizer. The winning contractor (Lockheed Martin) wanted to make all these structural elements primarily out of nonmetallic composite materials to lower weight and enhance stealthiness. However, experience suggested that finished load-bearing structural parts manufactured from traditional aerospace composite materials and processes generally averaged from $600 to $1000 per pound. Using these processes and materials could rapidly escalate the cost of JASSM past the target and even beyond the ceiling prices.

Lockheed Martin and its subcontractors began looking for solutions in the commercial world. Eventually engineers examined a process used for decades in making fiberglass hulls for pleasure boats called Vacuum-Assisted Resin-Transfer Molding (VARTM). This process produces finished fiber-composite parts that cost about $5 per pound. The resulting parts are not appropriate for aerospace applications. However, engineers experimented with variations on this process using different materials systems. Eventually an approach was discovered that, while more expensive than the VARTM process for boat hulls, turned out to cost only a fraction of the cost associated with traditional aerospace approaches that require high temperatures and pressures for curing, and thus need to be processed in expensive autoclaves. The modified VARTM approach was used for the body of JASSM. In addition, engineers developed a lower-cost automated braiding platform to lay down the fiber matrix for the body that was based on commercial machines used to braid socks, shoe laces, and freeway pillar reinforcement rings.

A similar approach of trying to find low-cost solutions from the commercial world was tried with the wings and vertical fin of JASSM. Lockheed Martin adopted a variation of the same process used by commercial firms to build surfboards and windmill blades for wind-driven electrical-energy generators. This process uses an outer composite shell and an inner foam core to form a durable, lightweight structure. Although the process had to be modified considerably, the contractor claims it resulted in a large savings compared with traditional aerospace composite structures costs.

In the case of the JASSM engine, Lockheed Martin used a combination of approaches. First, to save development costs on a new engine, designers selected an existing GOTS engine that had been used to power the Harpoon antiship missile for two decades. Second, the prime contractor helped the engine vendor lower the cost of the engine by one third by replacing outdated Mil-Spec parts and technology on the engine with modern but much less expensive commercial parts and technology. For example, the old Mil-Spec analog engine controller was replaced by a modern digital controller. This latter technology was based on an off-the-shelf antiskid processor used by the automobile industry. As with WCMD, many other automotive- and industrial-grade non–Mil-Spec parts were used. In general, the prime contractor asked the subcontractors and vendors to qualify the commercial parts if extra testing was needed.

In all these areas, the JASSM prime contractors used a common mechanism from the commercial world to keep costs under control: "must cost." Aggressive cost targets for each major subsystem and component such as the guidance and control units were provided to vendors. This in turn encouraged vendors to insert COTS parts and technology to keep costs down.

In cases where no existing commercial product existed to meet the need, JASSM engineers sought out existing military technologies and parts to avoid the expense of having to develop entirely new items. To achieve its performance requirements for autonomous terminal target acquisition and guidance, JASSM needed to use advanced sensors with target recognition capability. No appropriate commercial technologies existed to meet these needs.[25] However, according to some published sources, Lockheed Martin and its subcontractors were able to develop a derivative of the Imaging Infrared (IIR) seeker developed for the Hellfire and Javelin antitank missiles that is appropriate for JASSM. It is claimed that this seeker fills the basic requirement and costs only $50,000.[26]

---

[25]Early in the program, the U.S. GAO identified the automatic target recognition requirement and autonomous guidance system on JASSM as areas of high technical risk that could cause schedule slippage and cost growth (U.S. GAO, June 1996). More is said on this below.

[26]This seeker does not provide true all-weather capability. It is limited to a 1500-ft ceiling and 3-mi visibility. See *Aerospace Daily* (30 April 1998).

In a similar manner, Boeing, the losing contractor, worked closely with its subcontractors to reduce costs by adopting existing GOTS military hardware where commercial technology did not exist. In the case of the terminal guidance system, Boeing adopted a derivative of the infrared seeker already used in the AGM-130 powered standoff weapon. Instead of developing a new subsystem, Boeing incorporated the guidance system for its JASSM design that it was already using for its JDAM kit, and also used an antijamming GPS receiver already developed by the Air Force. Finally, Boeing's design also made use of the autonomous target recognition software that had already been developed for its Stand-Off Land Attack Missile Extended Response (SLAM-ER) missile under development for the Navy (Fulghum, 1998).

In summary, the focus on CAIV required a conscious effort to avoid gold plating and "requirements creep," the use of mission performance requirements, a heavy emphasis on "must cost" pricing, contractor configuration control, and Mil-Spec reform. This use of CAIV and a "must cost" commercial-like approach in turn encouraged contractors on these three munitions pilot programs to seek out commercial technologies and parts that could lower costs while maintaining adequate performance, or, if no commercial part existed, incorporate existing GOTS military parts and subsystems. Contractors were able to offer the government a richly varied menu of cost/benefit tradeoffs and alternative design solutions because the government provided no detailed system specification and did not demand the use of military specifications and standards. Indeed, the use of commercial parts, components, and processes was encouraged if it lowered costs and provided acceptable performance. The contractors were given almost total control over configuration, design, and technical solutions. If a commercial part slightly reduced environmental robustness, a contractor could still argue that the cost savings outweighed the loss in capability. The result was that the system design, its expected capabilities, the cost estimates, the technical solutions, the suggested parts and components, and so forth, all came from, and were "owned" by the contractors, not the government. Much lower costs than might be expected appear to have resulted from this approach. However, some doubts and potential problems remain.

More "Commercial-Like" R&D Program Structure

Another key aspect of approaches to military procurement that promote greater CMI is a more "commercial-like" R&D program with
more "commercial-like" contractor selection. The three munitions
programs under examination here focus on:

- Extended contractor competition during R&D

- Greater government-industry cooperation through IPTs and
  maximum sharing of information

- Past Performance Value (PPV) criteria and "rolling down-select."

Extended Competition. Many acquisition reform advocates clearly
perceive continuous and intense competition as the driving force
pushing firms in "textbook"-type commercial markets to lower their
prices, increase the quality of their products, and improve their
product performance. But in traditional military procurement programs, competition tends to last only during the initial concept development stage or the prototype demonstration/validation stage. In
these munitions pilot programs, reformers hoped to maintain competition longer.

Originally, acquisition officials had hoped to fund at least two competing contractors through the entire EMD program for at least one
of the munitions programs.[27] It rapidly became clear, however, that
this was not feasible from a cost standpoint. Instead, the officials
adopted the following approach. First, a considerable effort was
made to attract as many competitors as possible—particularly nontraditional commercial contractors—into the initial conceptual design stage. These contractors then took part in an initial, low-cost
paper competition. After completion of this phase of the competition, the EMD phase was divided into two parts. The first phase focused on lowering the technical risks associated with development
and manufacturing and reducing unit costs. The government funded
two competing contractors during this phase. At the end of the first
phase, the government selected one of the competing contractors to

---

[27]The R&D phases of all three munitions programs are essentially funded by traditional military cost-plus-fixed fee (CPFF) or cost-plus-incentive fee (CPIF) contracts.
More is said on this at the end of the chapter.

complete development. A major factor in the selection of the winner was the contractor's ability to achieve a low production price.

Thus, at least eight contractors submitted serious proposals in the original WCMD design phase, including companies that might not normally have entered a military system design competition of this sort. For JDAM, five contractors competed in the initial design competition before it was designated as a DAPP. JASSM received seven serious design proposals at the beginning of the program. With respect to WCMD and JASSM, the initial build-up period to a final RFP was characterized by intense cooperative interaction between the government program offices and each of the competing contractors regarding requirements, design concepts and approaches, and so forth.

Government-Industry Cooperation. Reformers also have observed that highly successful commercial firms often have relatively open and trusting relationships with key customers and suppliers. Information is shared; problems are worked out together. This contrasts sharply with the traditional adversarial relationship between the government and its contractors.[28] By bringing government and industry personnel together in IPTs and other cooperative arrangements, reformers hoped that program outcomes would be improved.

To support greater government-industry cooperation, all three munitions programs introduced the idea of multiple integrated government-contractor teams during the competitive EMD stage. The SPO established three separate IPTs for JDAM during Phase 1 EMD. Each contractor had its own exclusive government-industry IPT, whereas the SPO formed a third government-only core team. The two government-industry IPTs were walled off from each other and had no access to each other's data or documents (which were all treated as source-selection sensitive). Only the core SPO team had access to both government-contractor IPTs.

Most interestingly, the government members on the two contractor IPTs for JDAM were instructed to do as much as possible to help their

---

[28]Long-term cooperative relationships are also somewhat inconsistent with the "continuous and intense competition" that, as described above, was intended to be a key feature of these munitions pilot programs.

specific contractor win the competition. The government fielded teams of 10 to 12 military and civilian officials that "lived" at each contractor site, not to audit or check up on the contractor, but to help the contractor lower its costs and improve its approach. The contractors were allowed to use the government IPT members in any way they wanted. One contractor integrated the government members closely into its design and engineering groups, while the other used them more like consultants and advisors to clarify issues and problems. This concept was also meant to supplement the feedback provided by the periodic report cards issued by the core team during the rolling down-select process. In this way, the JDAM SPO hoped that both contractors would improve their proposals to such an extent during the EMD Phase 1 that it would be almost impossible to choose a winner.

Past Performance Criteria and Rolling Down-Select. Finally, reformers argue that contractors are motivated to perform well "in the commercial world" in part because they believe past performance helps to determine future success at winning contracts.[29] In contrast, government contracts in the past were generally awarded to firms whose proposals contained the lowest-cost estimates or promised the highest capabilities—with little regard for an individual firm's past record on delivering on promises. The PPV concept was developed to apply this commercial standard to the selection of military contractors.

The WCMD and JASSM competitions helped pioneer the concept of PPV as a criteria for contractor down-selects. In both WCMD and JASSM, past performance was assigned a weight equal to all other factors in the contractors' proposals for the down-select to the Program Definition and Risk Reduction (PDRR) phase. Perhaps most important, as part of the concept of a "rolling down-select," the contractors were informed of significant weaknesses and deficiencies in their proposals and their past performance evaluations. Contractors had full access to the criteria, standards, and methodology used in the evaluations. They also had numerous opportunities to respond

---

[29]Reputation is most important in markets for complex durable products, such as transport aircraft. However, it is becoming increasingly important even in markets for mass consumer products, such as televisions and refrigerators.

to and discuss government criticisms of both the technical proposals and past performance.

In the case of JASSM, officials developed broad categories of contractor past performance such as cost and schedule, product performance, and product reliability. Similar or related products developed and/or manufactured by the contractor in the past were examined. Thus, only past performance and capabilities of direct relevance to JASSM were assessed, such as aircraft integration, software development, and so forth. The outcome of this assessment was given equal weight to the content of the actual current proposal. In the JASSM system proposals, assessments of development and production costs were given equal weight with achievement of KPPs and other requirements. As a result of this process, Lockheed Martin and Boeing became the JASSM finalists in June 1996 for the PDRR phase.

After a similar process, Lockheed Martin and Alliant Techsystems won the WCMD first-phase contract in January 1995. But WCMD added an additional element to the down-select of its "Pilot Production" phase by conducting a live fly-off (or bomb-off?) using the two competing contractors' tail kits. The same F-16 carried one contractor's system on one wing and the competing contractor's system on the other, so that the exact same conditions would apply to both. This direct competitive fly-off helped lead to the selection of Lockheed Martin in January 1997 to conduct the next phase of the project.

In April 1994, before JDAM became a DAPP, program officials followed fairly conventional procedures to select Lockheed Martin and McDonnell Douglas (now Boeing) to continue competing during the 18-month Phase 1 EMD contract. As a DAPP, however, the JDAM program adopted the rolling down-select concept during its EMD Phase 1. The two contractors had asked for, and received, significantly different levels of funding, because they took different technical approaches. At first, the contractors were measured against their own SOWs rather than directly against each other. During the first year and a half, government officials provided the two competing contractors at six-month intervals with detailed "report cards" on their proposals, showing areas of strength and weakness. The contractors understood the measures of merit and had a full opportunity to respond to and even criticize the standards used if appropriate.

The central objective of this approach was to provide the contractors with as much leeway and as much information as possible, and let them compete against each other in a manner that mimicked what takes place in the commercial marketplace. In the end, it appears to have provided good results. The final source selection for JDAM EMD Phase 2 came down primarily to a question of production price commitments, and even then the decision was a close call.

Thus, as in markets involving generic goods and services, competition remained the central tool used to try to ensure low price and high quality in programs where traditional regulatory safeguards had been removed. Indeed, these programs took special measures to level the playing field and intensify the competition. It is noteworthy, however, that in all three programs established defense contractors were down-selected at relatively early phases of development. This suggests that DoD's efforts to encourage competition by lowering regulatory hurdles to the entry of firms without defense contracting experience may be offset by the new hurdle imposed by PPV criteria in down-selects. Government regulations provide that contractors having no relevant past performance are to be rated "neutral," putting them at a clear disadvantage relative to long-time defense contractors with strong past experience.

More "Commercial-Like" Purchase and Support of Developed Systems

To ensure the production of low-cost, high-quality, reliable and maintainable systems, the government developed a strategy to structure the purchase and support of the three munitions in a manner that would achieve the benefits enjoyed by buyers in routine commercial transactions. The main elements of this strategy are:

- Competitive fixed-price production commitments made during R&D

  — Applicable to initial production lots

  — Price reductions encouraged for later production lots through a "carrot-and-stick" incentive system

- System performance guarantees

- Full contractor responsibility for life-cycle reliability and "bumper-to-bumper" maintenance included in the system purchase price.

Government planners believed the bulk of the savings that would be generated by a more "commercial-like" acquisition approach on the three munitions pilot programs would accrue during the production phases. During the R&D phases, the government still paid up front for all costs, and indeed incurred extra costs by supporting two contractors during the first phase of R&D. However, the central focus of the R&D programs was to develop effective systems with much reduced production costs. For all three programs, the government initiated the R&D phase by providing the participating contractors with a production cost goal and cost ceiling beyond which the item would not be purchased. For all practical purposes, these goals and objectives were similar to airline "must cost" requirements placed on airframe prime contractors. The production price commitments provided by the munitions contractors, and the credibility of the estimates were central factors determining which contractor won the down-select at the end of the first phase of R&D. These prices tended to be far below the original government price goals. The problem for the government program managers then became how to ensure that contractors met the production price commitments and the performance and reliability guarantees.

For all three munitions programs, government officials have used procedures meant to emulate commercial-market approaches to guaranteeing production prices and system performance: Production Price Commitment Curves (PPCC) and warranties. The final contractor proposals for the second phase of R&D for these programs included fixed prices for low-rate production. The competing contractors for JDAM each agreed to submit an Average Unit Procurement Price Requirement (AUPPR) in FY93 dollars as part of the official System Specification they themselves wrote. The AUPPR had to include the cost of a full "bumper-to-bumper" warranty. The AUPPR applies to production lots 1 and 2, which make up the initial Low-Rate Initial Production (LRIP) phase. The system specification also included procurement price objectives for quantities in excess of 40,000 and 74,000 units, which in essence provided an estimate of the contractor's production learning curve. Thus, the contractors committed to a firm fixed price for the first LRIP lots at the beginning of full-scale development. Unlike customers in commercial commodity markets, the government required that cost data be submitted to back up the AUPPR. Nonetheless, the cost data requirements were simpler than those for a traditional program and were limited to "only" 15 pages.[30]

At the end of first phase of R&D, the contractors in all three munitions programs also provided a good-faith estimate of the production prices for production lots following LRIP. In the case of JDAM, the contractors provided nonbinding PPCCs for lots 3–5 (a total of about 8700 units) and agreed to submit PPCCs for lots 6–11 at the time of their lot 4 final price proposals. The government required no supportive cost data for these post-LRIP production PPCCs, but the contractors agreed to an extensive array of "carrot-and-stick" incentives to encourage their attainment. For example, should the contractor submit production prices when bidding for post-LRIP production lots that were at or below the original PPCC for lots 3–5, the contractor would enjoy the following benefits:

- The contractor remains the sole production source for an agreed number of lots, and the government will not request changes in subcontractors.

- The contractor retains full configuration control as long as changes do not reduce performance or affect safety of flight; however, changes must be documented and reported to the government.

- If the contractor is able to reduce its production costs through the insertion of new technologies or other efficiencies, such savings are retained entirely by the contractor as additional profit.

---

[30]It has been pointed out that the credibility of the AUPP estimates was built up over two to three years of working with government cost analysts and decisionmakers, and was supported by numerous affordability reports and briefings documenting specific design, manufacturing, management, and support concept changes. Note also that business relationships in many commercial industries—including the transport aircraft industry—do involve extensive sharing of cost data between buyers and suppliers. However, data-sharing in the commercial world is a product of mutually beneficial partnering arrangements, not TINA-type presumptions of attempted abuse.

- The contractor does not need to submit any type of cost or technical data to the government if performance, reliability, and delivery schedules are being met.

- The government will actively assist the contractor in reducing costs if requested, but will not pay to implement changes.

- There is no in-plant government oversight or inspection of the contractor or subcontractors and all acceptance testing is done by the contractor in accordance with mutually agreed upon procedures.

- The contractor receives an incentive fee if the accuracy and reliability of production units exceeds the specification.

Some analysts consider these "carrots" to be nearly revolutionary. Of particular importance is the profit incentive for the contractor to become more efficient and insert new technologies on his own initiative, aided by the elimination of the need to provide cost data to the government. This mechanism was intended to encourage the contractor to offer the lowest possible AUPP at the end of Phase 1 R&D. However, for the government to eventually enjoy some price benefit from contractors who continue reducing costs throughout production below the PPCC, the PPCC would have to be negotiated for later lots.

The munitions contracts also contained "sticks" to protect the government from unsatisfactory contractor performance, particularly in price and system performance. These measures can be implemented by the government if the contractor submits a price bid for a production contract lot that exceeds the PPCC for previously negotiated lots. However, there is a grace period during which the contractor can explain the reasons for exceeding the PPCC. If the government does not accept the explanation, the following measures can be taken:

- The contractor must submit fully compliant certified cost and pricing data in accordance with TINA and other regulations.

- The government may reestablish control over configuration.

- The contractor must prepare and provide a fully compliant Mil-Spec data package free of charge within one year.

- The contractor must fully qualify at his own expense and within 12 months a new contractor as a second source for production, where full qualification is defined as delivery and acceptance of a production unit by the second-source contractor and 10 success-ful flight tests.[31]

- The government may impose in-plant oversight and testing.

- The incentive fee option is eliminated.

Clearly, the most undesirable "stick" from the contractor's perspec-tive is the requirement to qualify a second source at his own expense. This stick is an attempt to simulate the incentives in the commercial world where, in most cases, an unsatisfied buyer has the option of turning to a competing supplier of the same or similar product. This option encourages the original supplier to fulfill his promises to the buyer. In the case of unique military hardware, especially when the government does not control the data package, the existence of other suppliers of nearly identical items is unlikely. Therefore, the contractor's penalty for failing to meet the promises to the govern-ment buyer on the three munitions programs is that the company must create a new supplier at its own expense—a severe penalty in-deed, and presumably a strong incentive to perform as promised.[32]

At least some JDAM contractor representatives view the reality somewhat differently. According to one contractor representative, in practice the formal incentives against price gouging become rela-tively weak by production lots 6–11, which represent the bulk of all production. From the contractor's perspective, a hard fixed-price commitment clearly exists for LRIP lots 1 and 2; and a somewhat softer commitment exists for lots 3–5. But in the contractor's view, it would be difficult for the government to enforce the requirement to qualify a second source if there are problems with the PPCC, particu-

---

[31]The government may impose fines of $20,000 for each working day up to a total of $5 million for failure to meet these requirements within schedule. Some of the terms may vary for each system or have been amended. For example, the WCMD contract apparently permits the contractor 18 months to qualify a second source.

[32]There is some question, however, as to the credibility of the threat to second source the hardware because of the proprietary nature of the data rights. Loss of reputation may be a more significant threat, especially given the increased importance of PPV in DoD source selection.

larly for lots 6–11, primarily because of issues related to proprietary data. According to another contractor representative, there really is no credible element among the contractual "sticks" that prevents price gouging in lots 6–12.

On the other hand, the contractor representatives say that there is a strong incentive to hold to the PPCC and not price gouge—reputation plus the trend toward using past performance in future contract awards. This incentive, they argue, works extremely well in the commercial aerospace world, and good faith and past performance are the keys to protecting the government from price gouging.

In addition to cost, government planners were concerned about system performance, including reliability. The government decided to provide three types of incentives to ensure that the contractor achieved system performance goals. These included:

- A commercial-style "bumper-to-bumper" warranty that includes system performance, reliability, and support

- Linking receipt of the PPCC incentive "carrots" to achievement of the performance specification

- The establishment of a formal dispute resolution process.

From the beginning of the program, both competing contractors accepted the concept of a commercial-style performance guarantee that requires that contractors meet all of the system specification requirements. The terms of the guarantee are flowed down to the major suppliers and vendors by the prime contractor. In the commercial transport industry, prime contractors often provide specific performance and reliability guarantees that entail cost penalties to the prime contractor if they are not achieved. Sometimes airlines try to negotiate the tradeoff of some warranties and performance guarantees for lower system prices. In the case of JDAM, there is no explicit cost penalty for not meeting specification requirements. However, unless all the contractor's kits meet the full specification requirement as determined by the government customer, the contractor does not enjoy the benefits of the PPCC "carrots," most particularly the promise not to "compete out" the production of the system to another source.

The JDAM warranty, which is similar to the warranties for the other two munitions programs, requires the contractor to replace or repair any JDAM kit that does not meet the system specification requirement or that contains defects in materials or workmanship, as determined by the government buyer. The warranty remains in force for 20 years if the kit remains in its shipping container and for five years outside of the container. If the kit is properly repacked in its container, the 20-year warranty goes back into effect. The warranty also applies to 50 hours of carriage life on the pylon of a combat aircraft, and includes a specific number of on-off operating cycles of the system during flight.

Many of the aspects of the warranty are similar to the standard commercial transport rules for Aircraft on Ground (AOG) resulting from a broken part. The warranty requires the contractor to ship out repaired or nondefective kits within a specific time period (within one business day for the early low-rate production lots). The contractor must pay for the cost of shipping to any place in the world. The warranty is not unconditional. It does not cover combat damage, uncontrollable events or misuse or abuse by the government.[33] On the other hand, in the case of JDAM, neither the contractor nor the government expect detailed records to be kept on specific kits that can prove how long the kit has been out of its container or how many hours it has flown on a pylon. In other words, implementation of the warranty is predicated on good-faith intent on both sides.

Nevertheless, all three munitions contracts also include provisions for a formal third-party dispute resolution process if the government and contractors disagree regarding the application of the warranty or other aspects of the contracts. The process entails the use of a Dispute Resolution Board (DRB) made up of three members who do not represent either party. Two of the members are chosen by each party from a list of five candidates provided by the other party. These two members choose the third member. Acceptance of a DRB finding is voluntary by both parties. However, all opinions and materials used in a DRB proceeding can be used in traditional dispute resolution procedures or in litigation.

---

[33]Past government attempts to take advantage of commercial-style warranties have failed because DoD could not prove that its usage patterns constituted normal and allowable wear and tear. See Kuenne et al. (1988).

Ultimately, the most important enforcement mechanism for the warranty is the same as in the commercial world: Reputation and past performance. If a contractor refuses to honor a warranty obligation that the government customer believes is clearly legitimate, this behavior will become part of the contractor's past performance record, which will be evaluated in competitions for future system development programs. This same incentive encourages companies in the commercial world to honor their performance claims and warranty commitments. Of course, for this incentive to be effective, there must be more than one credible source or contractor for future competitions, and past performance criteria must be important elements in down-selects.

In summary, these three munitions programs have been structured in a radically different manner from traditional programs in order to mimic the market incentives of the commercial world, promote insertion of commercial technology, and reap the claimed cost savings and efficiencies that are prevalent in the commercial marketplace. Although still in their early stages, the programs appear to be achieving many of the hoped-for benefits, namely, the rapid development of lower-cost, more-effective weapon systems.

MUNITIONS PILOT PROGRAM OUTCOMES

Program Cost

The three munitions programs were all structured to mimic the emerging "must cost" environment in the commercial transport sector. At least in the cases of JDAM and JASSM, the government customers established "must cost" maximum price thresholds above which the system would not be purchased. Later, contractors were encouraged through intense competition and the application of CAIV to develop aggressive price targets that were considerably below the maximum price thresholds. Finally, the contractors committed contractually to meet LRIP production price objectives, and accepted a series of "carrot-and-stick" contractual incentives to ensure the price goals would be met.

JDAM. As noted above, original government estimates for a JDAM-type weapon kit came to an AUPP for 40,000 units of $68,000 in 1993 dollars. After DoD designated JDAM as a DAPP, a "must cost" threshold of $40,000 or less per unit was established. Two additional pre-DAPP and post-DAPP estimates of JDAM program costs and AUPP are shown in Table 7.2. The data are from OUSD/AR's *1997 Compendium of Pilot Program Reports.* The first two columns are President's Budget (PB) actuals in then-year dollars. The second two columns are estimates in constant FY95 dollars projected by the Cost Analysis Improvement Group (CAIG).[34] The pre- and post-DAPP numbers from the President's Budget and the CAIG projections for comparable categories differ because of varying definitions and assumptions, and because the numbers were generated at slightly different times. Nonetheless, they both show a decline in AUPP for JDAM of at least 50 percent from the pre-acquisition reform numbers (columns one and three) to the post-acquisition reform numbers (columns two and four). Both sets of estimates show a 40–50 percent

Table 7.2

Pre- and Post-DAPP JDAM Program Costs and AUPP
(in $ millions except for AUPP)

Cost Element	PB FY95	PB FY98	CAIG I (FY95$)	CAIG II (FY95$)
R&D	549.7	462.9	346	380
Aircraft integration	TBD	TBD	893	478
Procurement	4874.9	2062.8[a]	3593	2012
O&S	TBD	TBD	290	130
Total cost	5558.8	2525.7	5122	3000
AUPP[b]	65.9	23.4	48.6	24.4

SOURCE: Based on data from OUSD/AR (1997b, p. 1–4).

NOTE: O&S: Operations & Support; TBD: To Be Determined.

[a]Assumes total procurement of 87,496 units. All other numbers assume total procurement of 74,000 units, except for the AUPP numbers, which assume 40,000 units. Current total production is expected to be 89,000 kits plus foreign sales.

[b]Thousands of dollars; 40,000 units.

---

[34]The CAIG resides within the Defense Department's Office of the Director, Program Analysis and Evaluation. The CAIG advises the Defense Acquisition Board on weapon system cost estimation, reviews, and presentation of cost analysis of future weapon systems. The CAIG also develops common cost-estimating procedures for DoD.

decline in total program costs for JDAM after acquisition reform. In addition, although not shown in exactly comparable terms, the AUPP numbers are considerably below the $40,000 "must cost" threshold established at the beginning of the program.

In contrast, JDAM R&D savings as shown in the *1997 Compendium of Pilot Program Reports* (Table 6) appear small or non-existent. Excluding aircraft integration costs, the PB numbers show about a 15-percent savings. However, the CAIG projections show an increase in R&D of about 10 percent. The CAIG numbers indicate an almost 50-percent decline in aircraft integration costs, but this improvement arose primarily from a reduction in the number of "threshold" aircraft requiring integration. On the positive side, the CAIG projections estimate a greater than 50-percent decrease in O&S costs.

Table 7.3 presents evidence from recent published sources at the U.S. Air Force Air Staff, using slightly different data. These data show a slightly smaller savings in development, at just under 15 percent. On the other hand, they indicate an even larger decline in AUPP to less than one half the cost in FY93 constant dollars, even when shown in then-year prices unadjusted for inflation.

The most recent published and unpublished sources suggest that by 1998 the AUPP for JDAM in FY93 dollars stood at around $15,000, and that the then-year dollar AUPP in FY98 stood at about $18,000. However, resolution of some technical problems that were detected in 1997 during development and testing may lead to a real increase of 4–5 percent in both development costs and AUPP. According to one published source, the added cost to the JDAM unit price in FY98 dollars is about $850 (*Aerospace Daily*, 26 August 1998). More is said on this below.

Table 7.3

Pre- and Post-DAPP JDAM Development Cost and AUPP

Cost Element	PB FY95	PB FY99
Development ($ million)	$549.7	$469.3
AUPP[a] ($ thousand)	$42.2 (FY93)	<$20

SOURCE: SAF/AQ (December 1997).

[a]40,000 units.

In summary, in constant FY93 dollars, the 1998 AUPP remains less than one half the procurement price estimated before the program became an acquisition reform pilot program. With a total buy now projected on the order of 89,000 units, this results in an inflation-adjusted procurement cost savings to the U.S. government of at least $2.0 billion.

WCMD. AUPP savings for WCMD on a percentage basis roughly equal those of JDAM when initial pre-reform estimates are compared with post-reform estimates. At the beginning of the program, the AUPP for WCMD had been projected at $25,000 in 1994 constant dollars for 40,000 units. This price included the Average Field Installation Unit Price, which covered contractor installation of the kit in the field. As of mid-1997, the 1994 constant dollar AUPP stood at $8937—a full 64 percent below the original "must cost" price (SAF/AQ, 1997).

In late 1996, with R&D for WCMD nearly complete, Air Force officials estimated a cost savings on EMD of 35 percent resulting from acquisition reform. This estimate was based on comparing the initial government estimate of supporting two contractors at a then-year dollar cost of $65.6 million compared with a projected total EMD then-year dollar cost of $42.9 million. Unfortunately, a year later in late 1997, several technical problems were identified during testing that required correction. The technical fixes led to a small increase in total EMD costs. Published sources claim, however, that the contractor agreed not to increase the AUPP (*Aerospace Daily*, 23 March 1998).

JASSM. Finally, although still in an early stage of development and experiencing some test problems, JASSM also appears to be fulfilling the promise of a more commercial-like acquisition approach by greatly surpassing its original goals for low-cost pricing. JASSM began with a must-not-exceed ceiling average unit price goal of $700,000 in FY95 constant dollars, and target price goal of $400,000 in FY95 constant dollars, for a production run of 2400. The $700,000 price ceiling goal was confirmed by a CAIG estimate. Government analysts estimated total development costs in FY95 constant dollars at $675 million.

In early April 1998, the Air Force down-selected to one contractor to complete development of JASSM. The winning contractor, Lockheed Martin, committed to an AUPP for the first 195 missiles of $275,000 in FY95 constant dollars, more than 30 percent below the target price of $400,000 and more than 60 percent below the threshold ceiling price of $700,000. Boeing, the losing contractor, also came in with an offer under the target price with an AUPP of $398,000 for lot 1.[35]

The development phase is also expected to cost approximately 30 percent less than original projections, and far less than the amount spent on the failed TSSAM program. Measured in FY95 constant dollars, the contracts awarded to the two contractors for the JASSM initial PDRR phase totaled $237.4 million (*Air Force News*, 1996). The full-scale development phase was expected to cost on the order of $200 million. However, a restructuring of the development schedule, as discussed below, has led to an estimated increase in EMD costs to about $240 million. This is still well below the original FY95 constant dollar projections of $675 million.

Performance and Schedule

Probably the single biggest concern of the opponents of a more "commercial-like" acquisition approach is that the elimination of regulatory safeguards and the insertion of commercial technologies into weapon systems will result in inadequate performance or performance shortfalls. For the most part, the three munitions pilot programs under consideration here do not indicate that these concerns are warranted, although certain technical difficulties have raised some red flags about the compatibility of commercial-like tradeoffs between flight safety and cost reduction and DoD's traditional desire to "gold plate" systems to ensure high margins of safety.

JDAM has experienced several high-visibility technical problems during its aircraft integration testing. Most of these have been solved without difficulty. For example, early in the flight test program problems were experienced with radio-frequency components and the

---

[35]See *Aerospace Daily* (30 April 1998). Recent accounts report the price has risen to $317,000 in FY95 constant dollars because of a decrease in the size of the initial buys. Yet this AUPP is still more than 20 percent below the original target price.

GPS systems. Later testing showed that the 2000-lb BLU-109 and the 1000-lb Mk-83 versions of JDAM were unstable at high angles of attack. This problem reduced the delivery envelopes for both weapons. To solve it required redesign of the aerodynamic strakes attached to the sides of the bomb as well as flight-control system software redesign and retesting (ODT&E, February 1998).

In addition during the JDAM flight test program, engineers found that unanticipated system vibration was causing problems in the transfer alignment of the inertial measurement unit (IMU). The problems arose only with the Mk-84 2000-lb variant of the JDAM kit and only when it was mounted on the inboard pylons of an F/A-18 Hornet operating at low altitudes and high speeds. This concurrent combination of kit type, aircraft type, mounting position, altitude, and speed is quite unlikely, especially given that the F/A-18's inboard pylon is typically used for fuel tanks, not weapons. Boeing had not designed JDAM for such a scenario. Nevertheless, Boeing was able to fix the problem by modifying the IMU's vibration isolator ring and sculling algorithm.

However, when the JDAM test units were then subjected to this high dynamic-load region for more-extended periods of time, it was found that the commercially derived friction brake could not withstand the unexpectedly high aerodynamic forces. The friction brake holds the fins steady prior to launch, so the result was fin and fin shaft fatigue from excessive vibration and movement. Once again, this caused problems in the transfer alignment of the IMU, and worse, caused fins to move or fin shafts to break prior to aircraft separation.

Boeing's initial attempts to solve the friction brake problem proved inadequate. Boeing engineers then adopted an entirely new approach based on a positive fin-locking mechanism that "nails-down" the fin until launch by inserting a metal pin into a hole in the fin. The pin retracts into the tail kit within one second when the JDAM-equipped bomb is dropped. In addition, the fin shafts and other parts had to be strengthened.

The additional nonrecurring engineering and the need for using more-expensive parts during production have resulted in a 4–5 percent increase in EMD costs and in AUPP, as mentioned above. This is not trivial with a buy of approximately 89,000 units; the additional procurement cost is on the order of $75 million or more. Nonetheless, the JDAM price is still well below the threshold and target prices established at the beginning of the program.

Were the JDAM technical problems caused by the use of commercial parts and technologies as part of CAIV? The direct answer appears to be no. Although the friction brake that proved inadequate was an inexpensive commercial derivative item, its inadequacy probably arose from Boeing's failure to calculate correctly the magnitude of the dynamic forces to which the JDAM Mk-84 tail fins would be subjected under certain special conditions. However, the problem occurred in part because Boeing placed a heavier emphasis on cost reduction than on designing for a low-probability worst-on-worst case scenario. Thus, it could be argued that the commercial-like approach taken by Boeing was incompatible with DoD's desire to achieve very high margins of safety.

Interestingly, both WCMD, developed by a different contractor, and the Joint Stand-Off Weapon (JSOW),[36] which is not an acquisition reform pilot program, experienced similar problems during development. During testing in late 1997, WCMD showed fin vibration and flutter problems when carried on an F-16 at supersonic speeds. Lockheed Martin engineers concluded that they had to use the same type of fix as Boeing engineers developed for JDAM: a positive fin-lock mechanism. The Air Force also encountered problems with the WCMD autopilot software during testing in late 1997. This problem was resolved fairly quickly (*Aerospace Daily*, 20 February 1998).

In most areas unaffected by technical developmental problems, JDAM and WCMD seem to have already met or exceeded their critical performance and reliability requirements. Probably the single most important requirement for these two weapons is accuracy. JDAM started with a 13-meter CEP requirement. During developmental and operational testing by the Air Force in late 1996 and early 1997, JDAM achieved an average CEP of 10.3 meters. By late 1998,

---

[36]Developed by Raytheon Texas Instruments, JSOW is a winged stand-off unpowered precision glide munition that comes in three variants and delivers unitary or submunition warheads of approximately 1000 lb. It has a range of 15 to 40 n mi depending on launch altitude. Like JDAM and WCMD, JSOW is guided by a GPS link and an onboard IMU. Like JASSM, a planned JSOW variant (AGM-154C) will have an IIR terminal seeker. The program is a joint Navy-Air Force program led by the Navy.

one source claimed that JDAM was achieving an average 9.7 CEP with an actual average miss distance of 6.5 meters (*Aerospace Daily*, 22 September 1998). Because of the success of the initial developmental tests, the Air Force authorized low-rate initial production in April 1997.

The true test for JDAM, however, came during the extended air campaign over Kosovo in early 1999. Between late March and early May, six B-2s delivered in excess of 500 JDAMs against targets in Kosovo—11 percent of the total bomb load dropped by U.S. forces during this period. Taking advantage of the GPS Aided Targeting Systems (GATS) on B-2s, JDAMs reportedly scored an average CEP of 6 meters, compared to the original 13 meter requirement.[37]

WCMD started with a threshold accuracy CEP requirement of 100 feet and a target CEP of 80 feet. WCMD has consistently achieved accuracies that greatly exceed the target CEP in developmental testing with launches at subsonic speeds. During testing in mid-1998, WCMD is reported to have achieved miss distances of 5 to 30 feet. It is for this reason that the Air Force approved low-rate initial production in August 1998 (*Aerospace Daily*, 4 August 1998). Once the fin-locking mechanism is installed in later production lots, accuracy with launches at supersonic speeds is expected to meet or exceed the initial requirement.

JDAM's technical problems have led to a restructuring of the developmental and operational test programs, the production program, and a delay of about a year in the procurement of the BLU-109 2000-lb bomb variant. In April 1997, the Air Force authorized LRIP of 900 JDAM kits for the Mk-84 bomb. Confidence in the weapon was so high that in 1997 Boeing delivered 140 "Early Operational Capability" JDAM kits to the operational B-2 wing at Whiteman Air Force Base. Originally the Air Force had planned to enter into full-rate production in 1998 with both the BLU-109 and Mk-84 2000-lb bomb kit variants. In late 1997, however, the Air Force delayed full-rate

---

[37]The most infamous example of JDAM's remarkable accuracy came when B-2–launched JDAMs precisely hit a building the heart of a dense urban area in Belgrade. Unfortunately, the U.S. government had misidentified the building. Instead of an important Serb target, it was the Chinese Embassy. See Bill Sweetman, "Coming to a Theatre Near You," *Interavia Business and Technology*, July 1999.

production and substituted a second lot of low-rate production made up exclusively of Mk-84 variants. The purpose of this change was to permit additional flight testing to work out the flight instability problems encountered with the Mk-83 and BLU-109 JDAM kits, and to continue development of the fin-locking mechanism necessary to qualify the Mk-84 for the F/A-18 inboard pylons. Air Force officials claim that this change will have little effect on the production program, since approximately the same number of kits in the same bomb size category will be procured as originally planned in 1998.[38] Because of the Kosovo air war, JDAM production was increased by 50 percent in May 1999.[39]

The WCMD technical problems led to a similar restructuring of the operational test and production phases of the program. The Air Force had originally planned to authorize LRIP in February 1988. Following the discovery of the supersonic launch-fin flutter and autopilot software problems in November 1997, the Air Force stopped operational testing and delayed LRIP until fixes could be found. However, it was determined to maintain the schedule for initial operational deliveries in July 1999. As a result, the Initial Operational Test and Evaluation (IOT&E) phase of the program was divided into two parts. The first part of the restructured IOT&E program tested subsonic launches from B-52s. These tests, which proved highly successful, permitted the authorization of LRIP in August 1998 and meant that initial operational deliveries to B-52 squadrons could take place early in 1999, three to five months ahead of schedule. The second part of the restructured IOT&E program flight-tested the fin-lock mechanism that had already been designed and ground-tested. Program officials claim that this restructuring had little effect on the production schedule. The only significant consequence, they argue, is that the money that was going to be used to incorporate a small electronics upgrade in the WCMD kit had to be spent on the software and fin-lock fixes (*Aerospace Daily*, 29 April 1998).

---

[38]See *Aerospace Daily* (15 December 1997 and 17 December 1997). The BLU-109 is designed to penetrate and destroy harder targets than the Mk-84, so some capability will be lost. Also, the first two LRIP production lots of Mk-84 JDAMs will not have the pin-locking mechanism fix, so they will not be usable on F/A-18s. Also see ODT&E (February 1998).

[39]See *Aerospace Daily* (3 May 1999).

The JASSM program is still in the early stages of R&D. GAO published a report on the JASSM program in 1996 that concluded that in the long run the risk of cost growth and schedule slippage was high (GAO, June 1996). This conclusion was based on the view that the JASSM development schedule was too short to permit maturation of the high-risk technical areas on the program—automatic target recognition, autonomous guidance, and aircraft integration.

Beginning in 1997, a variety of factors, including concerns over the level of technical risk remaining in the program, led to a restructuring of the program schedule. The original program schedule envisioned a 24-month PDRR phase beginning in June 1996, followed by a 32-month EMD phase beginning in June 1998. The nominal target date for the authorization of LRIP was April 2001. However, spurred by declining Navy interest in the program and significant congressional funding cuts in late 1997,[40] the Air Force restructured the PDRR phase. First, it was decided to down-select to one contractor on 1 April instead of in late June or early July at the planned beginning of EMD to save money. Second, the beginning of the PDRR phase was extended by about three months. Eventually, however, this evolved into a six-month extension or more, until November 1998. Thus, if one counts from the original contract award to the two contractors for the PDRR phase (June 1996), the PDRR phase has been extended by 25 percent over original estimates.[41] This length-

---

[40]The Navy was convinced that the Boeing Stand-off Land Attack Missile-Expanded Response (SLAM-ER), a modification of the existing Navy AGM-84 SLAM system (itself a modification of the Harpoon), would meet its requirements at less cost than the JASSM. Like JASSM, SLAM-ER is slated to have an automatic target acquisition system. It will have a >100 n mi stand-off range and deliver a 500-lb warhead. Congress authorized an analysis of which system best served both services' needs. A GAO study concluded that the JASSM potentially could be fielded earlier with superior capabilities and at less cost than the upgraded SLAM-ER Plus version, development of which would be necessary to meet all key JASSM performance objectives. OSD directed the Navy to maintain at least minimal participation in the JASSM program, but with the withdrawal of the F/A-18E/F as a "threshold" aircraft, active Navy participation essentially ended. See *Aerospace Daily* (29 September 1998).

[41]A contributing factor was the many weeks of delay the program experienced after the down-select to two contractors because of an official protest filed by Hughes, one of the contractors who lost in the first phase of the program. Because of the heavy use of "past performance" criteria by the government, all three of the munitions programs examined here filed formal protests after the initial down-select process. This led to considerable lost time and effort. However, in all cases, the government won its case against the protests.

ening of the PDRR phase provided more time for the contractor and the Air Force to reduce technical risk prior to full-scale development. In addition, technical risk was further reduced by eliminating some of the developmental tasks that had to be completed during the PDRR. For example, since the Navy had decreased its involvement in the program, the need to focus on early integration of JASSM with the F/A-18E/F fighter was eliminated.

In November 1998, press accounts reported that DoD also intended to restructure the full-scale development EMD phase by lengthening it considerably. According to these accounts, the EMD phase would be stretched from 34 months (originally 32 months) to 40 months, an increase of 25 percent, to further reduce technical risk prior to flight testing. According to a program official, "We [the Air Force and Lockheed Martin] decided that we needed to do more ground and captive-carry testing than we had planned in order to not have big surprises during the flight test program."[42] These schedule increases, officials predicted, would cause a commensurate increase in overall R&D costs.

At the time, these schedule extensions did not appear to be the product of major technical difficulties or problems caused by the innovative commercial approach but rather arose from a development schedule that the program director characterized as "unrealistic," given the level of technical risk involved. Even after the extension, the program director characterized it as "still the most aggressive new development for a weapon" in a long time.[43]

In April 1999, the first JASSM flight-test vehicle crashed, delaying the flight-test program at least a month. A "make-up" flight-test was scheduled for August. On August 12, Lockheed completed a successful separation and maneuver flight test.[44] Two weeks later, the Air Force announced a major restructuring of the EMD program. The Air Force and Lockheed agreed to delay the decision to begin LRIP by 10 months, from January to November 2001, to permit additional

---

[42]Terry Little, JASSM Program Director, quoted in *Aerospace Daily* (11 November 1998).

[43]*Ibid.*

[44]See *Jane's Defense News*, 19 August 1999.

flight tests of production-standard JASSM vehicles. The Air Force blamed technical problems and the contractor for the delays. According to press accounts, there were problems with engine development, the missile casing, and the air data system.[45]

There is no reason to believe that any of these developmental problems were related to the CMI approach adopted in the program. It is likely that such problems are common in the development of any complex new system. For example, JASSM's competitor, the SLAM-ER, failed its operational tests in August 1999, and as a result the Navy delayed full-scale production until at least the spring of 2000. SLAM-ER is usually considered a technologically lower-risk program than the JASSM because it is not a new development but a modification of the Harpoon/SLAM series of missiles.

The original 56-month development program has now been extended to 78 months, an increase of nearly 40 percent. Nonetheless, the new schedule is still well below the average munition development schedule of 110 months, according to the JASSM program director.[46] Assuming that the new development schedule can be met, JASSM will still be developed in less time than TSSAM. TSSAM was cancelled after about eight years of R&D, with development still incomplete. JASSM is now scheduled to be fully developed in six and a half years from program initiation, a schedule improvement over TSSAM of at least 18 percent.

THE NEED FOR GREATER RISK-SHARING IN ACQUISITION PILOT PROGRAMS

R&D Costs and Risks

Probably the least-imaginative aspect of the three munitions pilot programs under consideration here is that the government ended up negotiating what amounted to simple cost-plus R&D contracts[47]—

---

[45]*Aerospace Daily*, 30 August 1999.

[46]*Aerospace Daily*, 31 August 1999.

[47]For example, in the case of JDAM, the initial Engineering and Manufacturing Development (EMD-1) phase had a CPFF contract, whereas the EMD-2 phase had a CPIF contract.

fairly traditional military R&D contracts in which the government agrees to pay the contractor up front for essentially the entire cost of R&D before deciding whether to procure production items. These types of contracts place most of the financial risk on the government and can reduce the contractors' incentives to control costs. We believe higher-risk acquisition reform pilot programs need to be structured so that the R&D risk and costs can be shared by the contractor. This approach is necessary to help control R&D costs and production price in higher-risk programs.

As is discussed in Chapter Six, in the commercial aircraft sector, R&D risks and costs are increasingly shouldered by informal consortia of prime contractors and major subcontractors. Subcontractors become risk-sharing partners. Corporate investments in R&D often rise to the hundreds of millions and even billions of dollars with no certainty that a reasonable return on the investment can be earned through the sale of production items. This suggests that a greater degree of R&D risk-sharing by contractors is also possible in the military context provided that performance requirements are well defined.

The government considered requiring contractors to finance some portion of the R&D costs in at least one of the three munitions programs discussed here. This option was rejected. Government planners believed that no company would risk its own money developing an expensive high-technology item that had no realistic customer other than the U.S. Air Force or Navy. Rather, the antidote to the problem of contractor incentives in cost-plus contracts was seen as a strategy of maintaining competition as long as possible through R&D.

The reasoning went as follows. Contractors seek to win production contracts where the potential for profit is greatest. Thus, maintaining competition through R&D greatly increases government leverage over the contractors' performance. Program planners also reasoned that to bring new and innovative commercially oriented firms into the competition—firms that might not have the same financial clout as the defense industry giants—the government had to pay for R&D. The new contractors would not be familiar with the military environment or might be too small to risk self-financing a large R&D effort in an unfamiliar business area.

Two critical problems arose with the government strategy. First, inadequate budgetary resources existed to support more than one contractor during the critical second EMD phase. Indeed, in the case of JASSM, budgetary shortfalls necessitated the elimination of one of the competing contractors during the pre-EMD PDRR phase well before the beginning of full-scale development. In the case of WCMD, program officials had originally hoped to maintain competition into the second phase of R&D, but were unable to because of cost considerations. Second, acceptance of all R&D risk and cost by the government did not ultimately result in serious new competitors entering the market. Although multiple contractors competed for the first concept development phase of all three of the munitions pilot programs, in each case the two finalists were the same two military aerospace giants that dominate the aerospace sector: Boeing and Lockheed Martin.[48] The recent introduction of PPV criteria for source selection suggests that these two firms' dominance of military aerospace is likely to continue.

The commercial world clearly demonstrates that maintaining and enhancing competition among upper-tier contractors is a key element in ensuring lower-cost, high-quality weapon systems in a commercial-like environment. As discussed in Chapter Six, fear of losing a contract to either an existing competitor or a new entrant into the market is a powerful mechanism for motivating contractors to provide the lowest-cost, highest-quality product possible. Ironically, since the government paid for all R&D on most of the programs under consideration in this chapter, it became impossible to maintain competition into full-scale development. Furthermore, as demonstrated above, total government financing of R&D did not help new firms or commercial firms make it into the final stages of the competitions. To the contrary, defense contractors from the pre-acquisition reform era dominated all the competitions.

---

[48]The first-phase competitors for the munitions programs were: JDAM: Lockheed Martin, McDonnell Douglas (now Boeing), Raytheon, Rockwell (now Boeing), and Texas Instruments (now Raytheon); WCMD: Alliant, Boeing, Brunswick teamed with Rafael, Lockheed Martin, McDonnell Douglas (now Boeing), Rockwell (now Boeing), and Raytheon; and JASSM: Lockheed Martin, McDonnell Douglas (now Boeing), Hughes (now Raytheon), Texas Instruments (now Raytheon), and Raytheon teamed with Northrop Grumman.

Of course, there is still the problem of high market risk resulting from a single government buyer and the difficulty of selling military-unique items to other customers, which discourages firms from self-financing military R&D. There is significant merit to this argument, especially for large-scale, high-cost defense-unique items such as fighter aircraft. Nonetheless, as Chapter Six points out, the commercial transport market, where a few large consolidated airlines with similar requirements dominate the market, is less different from the military aerospace market than might initially be thought. Furthermore, even today the U.S government military market is far from monolithic. For many items there are significant separate military service markets. For example, the JASSM program has had to struggle to survive in the face of stiff competition both from the Navy-led SLAM-ER being developed by Boeing and from the JSOW under development by Raytheon Texas Instruments.[49]

In short, there is reason to believe that contractors will grudgingly accept more R&D cost and risk sharing, which will be necessary to maintain and enhance competition during R&D. On relatively low-cost, low-risk systems, or systems that are genuinely dual-use, contractors may be expected to finance R&D entirely on their own. For high-cost, high-risk, military-unique items, creative cost-sharing arrangements can be developed. Three existing programs illustrate creative ways in which this can be accomplished.

DarkStar, Global Hawk, and COSSI

Two programs initiated by DARPA, plus a modification and upgrade program administered by the services, illustrate some of the possibilities for increased R&D risk sharing between the government and defense contractors. DarkStar and Global Hawk, DARPA/Air Force programs for the development and possible production of HAE UAVs, each included terms that required the contractors to share in unanticipated R&D cost growth.[50] COSSI, a program leveraging commercial technology developments to reduce the operations and

---

[49]This competition has worked both ways. Raytheon and the Navy have had to dramatically reduce the projected unit price of the AGM-154C (Unitary) variant of JSOW in part because of price competition from JASSM.

[50]The DarkStar program was cancelled in early 1999.

support costs of legacy systems, requires contractors to share the costs of developing and testing a prototype ready for insertion in a military system.

The DarkStar Phase II (prototype) R&D baseline agreement is essentially a traditional CPFF/CPIF instrument. The government agreed to pay all Phase II R&D costs up to $115.7 million. The contractor could earn a relatively small fixed fee as well as a small incentive fee for meeting performance goals in four areas.[51] These fees would amount to about $8–$9 million or roughly 8 percent of R&D cost.

However, in a radical departure from traditional programs, the contractor agreed to pay 30 percent of Phase II R&D costs if they rose above $115.7 million, and 50 percent of R&D costs above $162 million. Further, the parties agreed to an absolute cap of $220 million on Phase II. Since relatively serious problems were encountered during the prototype flight-test program—resulting in a lengthening of the Phase II schedule—it is likely that the $220 million ceiling will be reached. If that is so, the contractor will be responsible for paying for nearly $43 million or more than 40 percent of a cost overrun of $104 million. This has been a painful experience for the prime contractor and a strong incentive to reduce technical risks and control costs in future phases of the program. It has saved the government a significant amount of money and made the contractor a risk-sharing partner in the development program, as is the case in the commercial world.[52]

The Global Hawk program has also experienced technical problems, cost growth, and schedule slippage during Phase II R&D. Program managers had originally planned to impose cost and performance discipline on the program by maintaining competition with at least two contractors throughout Phase II. However, funding shortfalls required an early down-select to one contractor. The Phase II agreement remained a traditional CPFF/CPIF instrument. As the result of significant cost growth, the parties renegotiated the Phase II

---

[51]The areas covered performance of the air vehicle (altitude and endurance), sensors (radar, electro/optical, IIR), and the command and control ground station.

[52]Although the contractors developed fixes for most of the developmental problems experienced during R&D, DoD decided to cancel DarkStar in early 1999 in order to focus on the potentially more operationally useful Global Hawk.

agreement in mid-1997. The new agreement resembles the DarkStar Phase II clauses that require the prime contractor to pay a percentage of cost overruns beyond a certain threshold and cap total government expenditure on the phase. The new Global Hawk program also requires major subcontractors to share in cost overruns, now typical procedure in the commercial aircraft industry.

In some ways, DoD's COSSI is even more innovative in cost sharing than the DARPA UAV/Air Force programs. COSSI projects are not subject to the normal DFARS regulations, functioning instead under the Other Transactions Authority for prototypes that is often used for DARPA programs. Whereas DarkStar and Global Hawk required contractor cost sharing only for cost overruns, COSSI requires contractors to share at least 25 percent of expected development costs.[53]

Although the basic technology for insertion under COSSI must be commercial (as broadly defined by DoD), it is recognized that significant Non-Recurring Engineering (NRE) is likely to be necessary both to adapt the commercial technology to the military system and to modify the military system to accept the commercial technology. In the last several years, COSSI has stimulated scores of proposals, many of them from nontraditional and commercial firms. On average, the firms have proposed that the government finance just 50 percent of the NRE. In some cases, contractors are paying up to 70 percent of the NRE. Interestingly, COSSI provides absolutely no guarantee that upon the completion of Phase I (NRE and operational testing) the participating military service will buy any of the kits for insertion into military systems (DoD, 31 August 1998).

COSSI programs tend to be small. The average Air Force program in FY97 was funded at about $6 million a year from the government for two years. Total annual COSSI funding has been on the order of only $100 million a year, covering about 30 projects, and much of this money has been cut as a result of the "Bosnia Tax."[54] This money

---

[53]Contractors obtaining government funds through COSSI are expected to sell the final product at a target price agreed upon at the beginning of prototype development. It is not clear how this will work in practice; no COSSI prototypes have yet entered full-scale production.

[54]Funding for COSSI projects is expected to be on the order of $90 million for FY00 (DoD, 2 February 1999).

goes almost entirely to the NRE and testing necessary to permit the commercial technology to be used in military-unique items. If the participating service does not then procure the technology, the contractor has few other potential customers for its militarized item. Thus, like the DarkStar and Global Hawk programs, COSSI seems to demonstrate that under certain conditions both commercial and defense contractors are willing to risk their own funds to finance military-specific R&D—even when there is no assurance that the government will procure the final product.

CONCLUSION

In our view the JDAM, JASSM, and WCMD munitions programs have gone a long way to demonstrate that a commercial-like "best" practice acquisition strategy can be applied to military-unique items, and that significant benefits, particularly in terms of production price, can be achieved. As in the world of commercial transports described in Chapter Six, a rigorously applied commercial-like "must cost" environment can produce dramatic results for the military. The experiences of DarkStar, Global Hawk, and COSSI suggest that greater risk-sharing between the government and contractors may also be possible, thus further reducing the costs to the government of weapon system acquisition.

In all three of the munitions programs, the likely acquisition prices appear to be considerably less than half of what they would be in a traditional military procurement program. With the large procurement numbers involved, this results in significant savings to the government. R&D costs so far appear to be running on the order of 20–30 percent less than traditional programs. Although R&D is not complete on any of these programs, and some technical problems have been encountered, operational performance capabilities appear on the whole to be meeting or exceeding original requirements. Some R&D schedule slippage has appeared, but all three of these programs began with aggressive schedules compared with traditional programs for similar products. Even with the schedule slippage, the R&D phase for all three has remained relatively short by traditional military developmental standards.

Key Elements Contributing to Success

In our judgment, the key aspects of the munitions pilot programs that have helped to achieve success are:

- Requirements reform and a closer customer-developer relationship through mechanisms such as IPTs, within a CAIV or "must cost" environment

- Contractor ownership of and responsibility for design, technical content, performance, reliability, and price.

In 1996, WCMD and JDAM program officials provided Air Force Materiel Command (AFMC) officials with a subjective percentage allocation of EMD and production savings accruing to their programs as the result of various acquisition reform initiatives (AFMC, 1996). In the EMD phase, 90 percent of acquisition reform savings were attributed to six features of the programs: CAIV (20 percent); IPTs (20 percent); performance objectives instead of specifications or SOOs vs. SOWs (25 percent); Total (Contractor) System Performance Responsibility (TSPR)[55] (20 percent); and insertion of nondevelopmental items and elimination of Mil-Specs (5 percent). For the much larger anticipated production savings, a full 90 percent was attributed to these same items plus contractor configuration control (10 percent), with the bulk of the savings attributed to CAIV (40 percent), TSPR (20 percent), and Mil-Spec elimination (20 percent).

One could quibble with specific percentages, but overall this assessment conforms with ours. The key elements are: (1) Making cost a fundamental system requirement ("must cost" in the commercial world), which is the objective of CAIV, and (2) granting system design authority and responsibility to the contractor, and the freedom to

---

[55]The "Acquisition Strategy" section of the AFMC *Guide to Acquisition Reform Cost . . .* (1996) explains that TSPR:

> . . . provides industry not only increased latitude in the design process for implementing system level solutions aimed at long-term sustainment, but provides clear accountability in design (CAID). Under TSPR the government continues to control system functional requirements while industry controls design/product requirements. Thus, the contractor is fully responsible for the integration of all systems, subsystems, components, government furnished property, contractor furnished equipment, and support equipment and must ensure no performance degradation after integration.

use that responsibility creatively. The latter element is characterized in the AFMC study by the concepts of performance objectives rather than technical specifications (SOO vs. SOW), contractor configuration control, and TSPR. Mil-Spec elimination and nondevelopmental items are enablers that permit contractors to seek out the lowest-cost, highest-leverage technologies whether they are in the commercial or military sector.

We believe it is crucial that future higher-risk, larger-scale acquisition reform pilot programs employ strong commercial-like "must cost" frameworks for the design, engineering, and development phases, and commercial pricing approaches for the production and procurement phases. A key element of each of the three munitions pilot programs is a strong focus on lowering production price, with "must cost" production price targets included in the operational requirement.

However, we believe that, as in the commercial world, contractors can take on greater price risk in the production phase as well as cost risk in the R&D phase. All the munitions programs assumed that the government should pay for all or most of the higher nonrecurring costs at the beginning of the program as the contractor moved down the production learning curve. This is why the concepts of AUPP and PPCC were developed. The AUPPR and AUPP represent average actual recurring costs for relatively small lots of very large production programs, after the nonrecurring costs have been paid by the government through a CPFF/CPIF-type contract. In contrast, commercial transport aircraft developers price even their first aircraft according to a projected average recurring and nonrecurring cost over a relatively large production run, even though they have no guarantee that they will sell any aircraft at all. This is because customer airlines in the commercial marketplace would not tolerate paying the high price necessary to cover the actual recurring costs to the manufacturer of the early production aircraft, which are high up on the learning curve.

Thus, a commercial aircraft developer may price his aircraft so that his financial breakeven point, where he begins to make a profit, does not come until many hundreds of aircraft have been sold. This approach encourages the manufacturer to continue every effort to reduce production costs, maintain high quality, and remain responsive to customer needs. Furthermore, it imposes greater discipline on the must-cost aspects of the design and development stage.

Defense contractors are beginning to consider commercial pricing approaches. There are some indications that Lockheed Martin adopted a modified commercial pricing approach for JASSM, by charging a price to the government for the first LRIP lot that is less than the company's actual production costs for that lot. On the C-130J Hercules program, Lockheed Martin has offered the government a commercial price alternative based on the assumption of significant sales over time, which requires the contractor to bear a significant amount of the cost risk during production.

A commercial pricing strategy combined with a fixed-price long-term contractor logistics support agreement based on mission performance goals could in principle further motivate contractors to offer lower fixed commercial-like production prices, insert cost-saving new technologies and processes into the production line, and enhance system reliability. This approach was discussed in Chapter Five. However, little experience with this approach has been gained on the munitions programs under consideration in this chapter, since they are to receive virtually no active maintenance during their storage lives. In addition, support contracts have not been negotiated for the Office of Technology Assessment (OTA) UAV programs. However, several contractors have negotiated interesting TSPR support contracts with the government on such aircraft as the F-117 and the C-17, which warrant further study.

Applicability of Lessons Learned

There are, however, legitimate concerns about some aspects of these munitions programs that suggest that the lessons learned may have limited applicability. First, all three programs aim at the development of single-use, unmanned systems. Reliability and maintenance concerns for multiple-use manned systems are therefore not addressed. Second, they are unusual in that they represent military-unique items that are intended for production and procurement in numbers that are very large by military standards.[56] It could be argued that the bulk of the overall program cost savings arises from the large number of items procured, so that the experience of these pilot programs may not be applicable to more traditional military procurement programs that have—by commercial standards—low procurement numbers. Third, some have considered these pilot programs to be relatively low-risk technologically. Indeed, candidates for DAPPs were required to be "low risk" to qualify for consideration (DoD, June 1998).

The third point can be disputed, particularly in the case of JASSM, which is not an official DAPP but contains most of the program elements of a DAPP. With its long and difficult development history, JASSM's predecessor, TSSAM, clearly demonstrated the technological complexity and risk inherent in developing a stealthy long-range cruise missile. In the case of JDAM and WCMD, both contractors dispute the alleged low-risk nature of the developmental programs. The contractors argue that even the integration of OTS subsystems into a new system, as well as the integration of the tail kits with the host aircraft, is technologically challenging. The technological problems already encountered on the JDAM and WCMD developmental programs, particularly with aircraft integration, seem to confirm that the development and integration of military-unique weapon systems is never without risk. Even in a fully commercial-like environment, realistic schedules must be developed to take into account the inevitable technical risk inherent in such systems.

In sum, we believe that a commercial-like acquisition approach as defined here could bring significant benefits to major Air Force acquisition programs, including those that entail much higher technical risk and the development of manned combat aircraft or other reusable systems with relatively lower production numbers. We recommend that DoD seek to expand the DAPP effort to include such programs. The JSF program has already made extensive use of CAIV and IPTs during its early phases. JSF would be an excellent candidate pilot program for application of the full panoply of acquisition reform measures during EMD. Based on our analysis of the commer-

---

[56]Anticipated production buys are: JDAM, 89,000 units; WCMD, 40,000 units; and JASSM, 24,000 units.

cial aerospace industry in the previous chapter, and the experience of DarkStar, Global Hawk, and COSSI, we recommend that future programs be structured to include greater risk-sharing between contractors and the government.

Chapter Eight

SUMMARY CONCLUSIONS

To achieve the benefits of CMI, advocates call for a relaxation of the regulatory restrictions that segregate weapon system acquisition from common commercial market practice and impose a regulatory cost premium on items purchased by the government. They base their arguments on two types of assumptions. First, they assume that an extensive "dual-use" overlap between commercial and military process and product technologies has created the potential for significant economies of scope and scale. Second, they assume that commercial business practices, together with the incentives and constraints provided by a commercial-like market structure, will spur the development of high-performing weapon systems at lower cost than can be achieved under the current heavily regulated military acquisition process.

However, although advocates claim that these barriers are preventing the acquisition process from operating more efficiently, critics argue that significant government regulation is necessary in a system that is fundamentally unlike the commercial world. In the view of the critics, highly specialized U.S. military requirements mean there is still relatively little dual-use overlap between commercial and military products and processes. Diversification of market and technical risk is not possible, so the existence of a specialized cadre of defense-oriented firms—with full government financing of military R&D—is unavoidable. Private-sector control over configuration management is also not practical because, for most weapon systems programs, private firms do not have the necessary incentives or information to make appropriate cost-performance tradeoffs. This implies that continued close governmental direction of product development is necessary. In this environment, complete elimination of such features as Mil-Specs, detailed contract requirements, and extensive governmental oversight removes necessary protections against the waste, fraud, and abuse of taxpayer money.

POTENTIAL FOR INTEGRATION OF COMMERCIAL AND MILITARY TECHNOLOGIES

In this study, we have closely examined the claims and counter-claims about the nature of the dual-use overlap between civilian and military product and process technologies in the context of radar-related and other RF/microwave devices. The questions we addressed were:

- Is the commercial market in military-relevant electronics large enough to encompass an adequate range of technologies, parts, and components required to support a comprehensive CMI strategy for military-specific microwave subsystems such as fire-control radars?

- Is the market driving technology at a rate and in a direction that meets national security requirements? In other words, can CMI provide the necessary and desired performance capabilities?

- Are there cost and schedule benefits from inserting commercially derived parts and technology into military systems such as RF/microwave systems?

Our conclusion is that, in defense-related microwave and millimeter-wave technologies, the promise of CMI and other acquisition reform measures is likely to be realized. In response to our first question, we found that:

- The commercial marketplace does appear to be increasingly driving the development of new technologies and lower-cost manufacturing processes in RF/microwave products relevant to defense applications. Commercial demand for sophisticated RF/microwave parts and devices is likely to far outstrip military demand in the next few years, and commercial design methodologies and process technologies are becoming increasingly relevant to military radar system design and development and other military microwave systems. Commercially developed RF/microwave parts and components are also becoming increasingly available for incorporation into military systems.

Thus, the technological breadth and depth in the commercial RF/microwave market necessary to support a comprehensive CMI strategy appears to be emerging. In fact, shrinkage of the military supplier base, together with the problem of increasing military parts obsolescence, indicate that the use of commercial-grade parts will increase dramatically whether or not a comprehensive CMI strategy is already in place.

We also conclude that CMI is likely to provide the necessary and desired performance capabilities, offering significant cost and schedule benefits from inserting commercially derived parts and technology into military systems. In response to our second and third questions, we found that:

- Commercially derived designs, technologies, and processes can be successfully applied to military RF/microwave systems with the potential of increasing performance. However, commercial technologies are probably not as relevant to the most advanced high-performance fire-control radars used in modern fighter jets, and some legitimate concerns remain about the long-term reliability and durability of commercial-grade parts and components.

- Evidence suggests that the systematic insertion of commercial parts, technologies, and manufacturing processes, combined with dual-use automated manufacturing, may reduce the costs of typical military digital avionics modules by 50 percent or more.

Finally, an important observation that emerged from our initial case study analysis was that:

- Effective implementation of a comprehensive CMI strategy may require granting configuration control and change authority to contractors during R&D and production, and—perhaps—throughout the life-cycle of a weapon system. This raises the potential for a fundamental change in the role of the contractor and the current military depot system. More analysis of this question needs to be undertaken.

MECHANISMS TO MINIMIZE THE RISKS OF POOR PERFORMANCE AND HIGH COST

The second set of questions addressed in this study focused on the risks DoD might face by shifting from reliance on regulatory constraints to increased use of commercial-market mechanisms to ensure access to high-performing, low-cost weapon systems. Specifically, the questions we addressed were:

- What mechanisms have commercial-market participants evolved to reduce risks associated with the development, production, and maintenance of large transport aircraft? To what extent are they relevant to DoD?

- To what extent, and with what success, have commercial-like approaches based on market mechanisms been applied to military programs, and what can be learned from them for future efforts?

After examining changes in the commercial aircraft industry as a result of airline deregulation, we conclude that current DoD policy on procurement of "commercial items" does not reflect the true variety and complexity of commercial buyer-supplier relationships and contract arrangements. There are many risk-management strategies now prevalent in the commercial aircraft industry that could be relevant to DoD. Examples include IPTs (encompassing users, buyers, system integrators, and vendors), "best value" sourcing through preferred providers, and various information-sharing and risk-sharing arrangements between buyers and suppliers. Probably the most important—and most currently relevant—strategy of all is the adoption of a "must cost" approach to pricing, which is the commercial aircraft industry's rendition of CAIV. However to make "must cost" work, airframers are giving their suppliers greater control over product configuration and design than they have before.

If DoD's approach to weapon system acquisition begins to resemble the approach used by commercial airlines to purchase airliners, DoD may expect to see an acceleration of the following trends:

- Greater emphasis by contractors on lowering the cost of purchasing and operating military aircraft as opposed to improving their performance characteristics

- Greater market and technical risk-sharing between prime contractors and suppliers of military aircraft systems, subsystems, parts, and components

- More intense competition between prime contractors accompanied by increased industry consolidation and greater foreign participation at all levels of the industry supply chain

- Greater integration of military aircraft R&D with maintenance, repair, and overhaul activities.

Will these trends prove beneficial to DoD? We believe the answer is a qualified "yes." If military contractors follow the precedents set by their commercial counterparts, we expect to see a decline in the cost to DoD of developing, producing and maintaining military aircraft. We caution, however, that cost declines may be accompanied by a diminution of the technical virtuosity of U.S. military aircraft if too much emphasis is placed on cost control relative to performance innovations. The "commercial approach" to minimizing technical risk is to limit performance innovations to those that are highly incremental.

Finally, recognizing that there are still important differences between commercial and military aircraft markets, we examined DoD's own initial experience in a variety of ongoing pilot programs aimed at testing a commercial-like approach to acquisition. A careful examination of three munitions acquisition reform pilot programs (JDAM, WCMD, and JASSM), as well as three other innovative acquisition reform efforts (DarkStar, Global Hawk, and COSSI) suggests that many acquisition reform measures have real merit, and that greater commercial-military integration is possible. In all three munitions programs, the likely acquisition prices appear to be considerably less than half of what they would have been in a traditional military procurement program. Further, although R&D is not complete on all of the munitions programs—and some technical problems have been encountered—operational performance capabilities appear on the whole to be meeting or exceeding original requirements. The R&D

phase for all three has also remained relatively short by traditional military developmental standards.

In our judgment, the main benefits of CMI for these acquisition reform pilot programs have not come from insertion of commercial technologies or the use of dual-use production facilities. For the most part, on all these programs the direct insertion of commercial technologies has been limited to the parts level. With the possible exception of the COSSI program, there are few cases where a major subsystem or component was designed or manufactured at a commercial facility. In no case did a traditionally nonmilitary contractor successfully develop a military-unique system.

The main benefits have come from the structuring and management of these programs in a manner that makes them more like complex commercial product markets where buyers and sellers establish and achieve price and performance targets in a cooperative rather than adversarial environment. Thus, the key aspects of these munitions pilot programs that have helped to achieve success are:

- Requirements reform and a closer customer-developer relationship through mechanisms such as IPTs and PPV criteria for "best value" sourcing, all within a CAIV or "must cost" environment

- Contractor ownership of and responsibility for design, technical content, performance, reliability, and price.

The real promise of CMI, therefore, is to help insert the incentives for price discipline and high performance that usually predominate in the commercial marketplace into the world of military R&D and production. One key to doing so successfully is to establish strict "must cost" guidelines while offering contractors real flexibility in design and technical content. A second key is to adopt mechanisms such as IPTs and "best value" sourcing to establish close working relationships with highly qualified contractors.

We close by cautioning once again that a CMI acquisition strategy does hold potential for a loss of useful military capability. To the extent that the old flexible budget approach to acquisition resulted in weapon systems with many unnecessary features ("gold-plating"), DoD's move toward a more commercial-like "must cost" approach to military R&D represents a transition to more responsible, cost–

effective stewardship of the nation's resources. However, to avoid accidentally sacrificing features that may prove to be crucial to successful mission performance, DoD must thoroughly understand the performance, schedule, and cost priorities for each new weapon system. This can be a daunting task in a global environment of great uncertainty about potential future threats, especially when weapon systems are expected to remain in active service inventories for decades. Perhaps no less important, DoD must be able effectively to communicate those priorities to the weapon system developers who are responsible for making the tradeoffs between them.

For these reasons, the issues of ensuring open-system architecture, developing a strategy of "continuous insertion" of new technologies, and providing incentives to contractors to continue to improve capabilities and reduce costs after production is under way deserve considerable further examination and analysis.

REFERENCES

Aerospace Daily, "Approval of Extended JASSM EMD Program Seen Imminent," 11 November 1998.

Aerospace Daily, "Boeing Presses 500-Pound JDAM Kit for U.S., International Buyers," 22 September 1998.

Aerospace Daily, "First WCMD B-52 Test Prepared for Continued Development Testing," 29 April 1998.

Aerospace Daily, "GAO Finds No Reason to Terminate JASSM," 29 September 1998.

Aerospace Daily, "JASSM Beats Cost Target by 40 Percent," 30 April 1998.

Aerospace Daily, "Navy Wants Upgraded JDAM for No More than $50,000," 26 August 1998.

Aerospace Daily, "Problems Force Delay in JDAM Full-Rate Production," 15 December 1997.

Aerospace Daily, "USAF Approves WCMD for Low Rate Production," 4 August 1998.

Aerospace Daily, "USAF Has Fix for One WCMD Problem," 20 February 1998.

Aerospace Daily, "USAF to Begin Planning JASSM Upgrades," 29 April 1998.

Aerospace Daily, "USAF Will Buy Only Mk84 JDAMs This Year," 17 December 1997.

Aerospace Daily, "Wind Corrected Munitions Dispenser Price Holds Despite Fixes," 23 March 1998.

Aerospace Industries Association of America (AIA), *Aerospace Facts and Figures*, Washington, D.C., various issues.

Air Command and Staff College, ACSC-1645-80 (DTIC #AD-B049 100), "Commercial Acquisition—It's Worth the Investment," 1980.

Air Force Materiel Command (AFMC), *Acquisition Reform Cost Savings and Cost Avoidance: A Compilation of Cost Savings and Cost Avoidance Resulting from Implementing Acquisition Reform Initiatives*, Draft report, 19 December 1996.

Air Force News, "Competing JASSM Contractors Chosen," June 1996.

Airbus Industrie, "1998 Annual Review," Webpage, January 1999, available at www.airbus.com/news_feature.

Alic, John A., Lewis M. Branscomb, Harvey Brooks, Ashton B. Carter, and Gerald L. Epstein, *Beyond Spinoff: Military and Commercial Technologies in a Changing World*, Harvard Business School Press, Boston, Massachusetts, 1992.

American Defense Preparedness Association, *Doing Business With DoD—The Cost Premium*, Arlington, Virginia, 1992.

Apple Computer, Inc., "Apple Accelerates Marketshare Strategy, Rolls Out Competitively Priced Macintosh Models," 21 October 1993.

ARINC, "ARINC Characteristics and Document Ordering," Webpage, 1998, available at www.arinc.com/Ind_Govt_Srv/Characteristics/charac.html.

Arnold, Wayne, "Mobile-Telephone Customers Face a Whirlwind of Jargon," *The Wall Street Journal Interactive Edition*, 16 September 1996.

Arpey, Gerard J., "The Challenge of Airline Finance," *Handbook of Airline Economics*, McGraw-Hill, New York, 1995.

Assistant Secretary of the Air Force (Acquisition), *JDAM—The Value of Acquisition Streamlining*, no date.

Aviation Week and Space Technology (AWST), "Airlines Urge 'No-Frills' Jets," 12 October 1998, p. 40.

Aviation Week and Space Technology (AWST), "Three Challenges Facing MRO Today," Advertiser-sponsored supplement, 7 July 1997, p. S1.

Berghell, A. B., *Production Engineering in the Aircraft Industry*, McGraw-Hill, New York, 1944.

Bierman, Howard, "Microwave and mm-Wave Technology: The Brains Behind Smart Weapons," *Microwave Journal*, June 1991.

Bilstein, Roger E., *The American Aerospace Industry: From Workshop to Global Enterprise*, Twayne Publishers, New York, 1996.

Boeing Company, "Boeing 717-200 Designed for Short-Range Market," Webpage, www.boeing.com/commercial/717/indepth. html. 1996–1998b.

Boeing Company, "Boeing 777 Background Information," Webpage, available at www.boeing.com/commercial/777family.   1996–1998a.

Buderi, Robert, *Echoes of A Wartime Campaign: The Invention That Changed the World; How a Small Group of Radar Pioneers Won the Second World War and Launched a Technological Revolution*, Simon and Schuster, New York, 1996.

Camm, Frank, *Expanding Private Production of Defense Services*, RAND, MR-734-CRMAF, 1996.

Canaan Group, "Aviation Maintenance in the Information Age," *Commentary*, Spring 1998.

Canaan Group, "The Changing Character of Aviation Maintenance: From Technical Imperative to Business Enterprise," *Commentary*, March 1996.

Canaan Group, "Parts, Partnerships and Technology: The Aircraft Component Overhaul and Repair Market in Transition," *Commentary*, May 1994.

Carley, William M., "How Honeywell Beat Litton to Dominate Navigation Gear Field," *The Wall Street Journal*, 20 September 1996, p. 1.

Carnegie Commission on Science, Technology and Government, *A Radical Reform of the Defense Acquisition System*, New York, 1 December 1992.

Center for Strategic and International Studies (CSIS), *Integrating Commercial and Military Technologies for National Strength: Report of the CSIS Steering Committee on Security and Technology*, Center for Strategic and International Studies, Washington, D.C., March 1991.

CFM International, "International Excellence," Webpage, 6 April 1998, available at http://www.cfm56.com/home.htm.

Chapman, Suzanne, "JASSM Competitors Chosen," *Air Force Magazine*, August 1996.

Charles, Richard, and Atef Ghobrial, "An Assessment of the Changes and Performance of the Aviation Industry," *Handbook of Airline Economics*, McGraw-Hill, New York, 1995.

Chenoweth, Mary E., Brent R. Keltner, and Ellen M. Pint, *Commerical-Type Contracting for Depot Maintenance Services: Findings on Policy Issues That Emerge from Air Force Experience*, Santa Monica, CA: RAND, AB-233-AF, August 1998.

Cook, Nick, and Duncan Macrae, "High Stakes in Aerostructures," *Interavia*, March 1997.

Coopers & Lybrand/TASC (C&L/TASC), "The DoD Regulatory Cost Premium: A Quantitative Assessment," annotated briefing prepared for Secretary of Defense William Perry, December 1994.

Defense Advanced Research Projects Agency (DARPA), *Rapid Prototyping of Application-Specific Signal Processors (RASSP) Program Architecture Guide*, Revision C, 14 April 1995.

Defense Science Board (DSB), *Report of the Defense Science Board Task Force on Defense Acquisition Reform*, DTIC-AD-A268734, July 1993.

Defense Systems Management College, *Commercial Practices for Defense Acquisition Guidebook*, January 1992.

Donlin, Noel E., *The Reliability of Plastic Encapsulated Microcircuits and Hermetically Sealed Microcircuits in MICOM Missiles*, Zeus Electronics, Concord, North Carolina, February 1995.

Drezner, Jeffrey, and Geoffrey Sommer, *Innovative Management in the DARPA High-Altitude Endurance Unmanned Aerial Vehicle Program: Phase II Experience*, RAND, MR-1054-DARPA, 1999.

Economist, "Airbus and Boeing: Deflating the Jumbo," 25 January 1997, pp. 58–59.

Economist, "Science and Technology: The Final Frontier," 27 July 1996.

Eisman, Melvin, and Hyman Shulman, "Fighter Aircraft Avionics: Cost Decomposition and Cost Drivers Assessment," RAND, unpublished briefing, March 1996.

Electronic Industries Association (EIA), *Results from the Workshop on Commercial-Military Integration in the Defense Electronics Industrial Base*, January 1997.

Federal Acquisition Institute, *Guide to FAR Changes Pursuant to FASA and Other Acquisition Reforms*, 18 October 1995. Available on the web at www.gsa.gov/staff/v/training.htm.

Federal Aviation Administration (FAA), *90-Day Report on Airline Safety and Security*, 16 September 1996. Available on the web at www.faa.gov/apa/publicat.htm.

Flanigan, James, "Gaining Altitude," *Los Angeles Times*, 28 February 1999, p. C-1.

Flight International, "IAI Seeks to Wrap Up FedEx Deal with Further Airtruck Investors," 26 August–1 September 1998, p. 13.

Frost and Sullivan, *World Commercial Avionics Market,* Research Report No. 5331-22, Mountain View, California, February 1996.

Fulghum, David A., "Boeing Unveils Stealth Standoff Missile Design," *Aviation Week and Space Technology,* 9 March 1998.

Galipault, John B., "Two If by Land; Three or More If by Sea," *Aviation Safety Monitor,* The Aviation Safety Institute, November 1991.

Gansler, Jacques S., *Affording Defense,* MIT Press, Cambridge, MA, 1989.

Gansler, Jacques S., *Defense Conversion: Transforming the Arsenal of Democracy,* MIT Press, Cambridge, MA, 1995.

Gordon, Robert J., *The Measurement of Durable Goods Prices,* The University of Chicago Press, Chicago, 1990.

Hardy, Quentin, "The Wireless Bloodbath: It's PCS Versus Cellular," *The Wall Street Journal Interactive Edition,* 16 September 1996.

Heppenheimer, T. A., *Turbulent Skies: The History of Commercial Aviation,* John Wiley & Sons, Inc., New York, 1995.

Holley, Irving B., *Ideas and Weapons,* New Haven: Yale University, 1953 (reissued by the Office of Air Force History, 1983).

Honeywell, *Commercial Aviation Systems Overview,* presentation at the Teledyne Controls Users Conference, 17–21 March 1997, Publication 41-7835–00-01, January 1997.

Honeywell, "Honeywell Appeals European Commission Decision on French Support for Avionics System," press release, 7 April 1998, available on the web at www.sac.honeywell.com/atsrel98. html#eu-7.

Honeywell, "Honeywell Supplies 'Via 2000' Avionics System for Boeing 717 Advanced Flight Deck," press release, 7 September 1998, available on the web at www.sac.honeywell.com/atsrel98. html#717_via.

Hostetler, Michele, "Untethered Data Dispatches Using Wireless Networking," *Investors Business Daily,* 20 August 1996.

Irving, Clive, *Wide-Body: The Triumph of the 747*, William Morrow and Company, New York, 1993.

Jane's Defense News, 19 August 1999.

John Deere & Company, "Sourcing Strategies," "Supplier Relationships," and "Cost Management Strategies," unpublished corporate documents, undated.

Jordan, William A., *Airline Regulation in America*, The Johns Hopkins Press, Baltimore, Maryland, 1970.

Kaminski, Paul G., "Affordable Radar Technology: The Defense Perspective," address to the 1996 IEEE National Radar Conference, Ann Arbor, Michigan, 14 May 1996.

Kaplan, Daniel, "The Pursuit of Competition: A Review of U.S. Public Policy in the Airline Industry," *Handbook of Airline Economics*, McGraw-Hill, New York, 1995.

Keating, Edward G., *Government Contracting Options: A Model and Application*, RAND, MR-693-AF, 1996.

Keating, Edward G., Frank Camm, and Christopher Hanks, *Sourcing Decisions for Air Force Support Services: Current and Historical Patterns*, RAND, DB-193-AF, 1996.

Kennedy, Michael, Susan Resetar, and Nicole DeHoratius, *Holding the Lead: Sustaining a Viable U.S. Military Fixed-Wing Aeronautical R&D Industrial Base*, RAND, MR-777-AF, 1996.

Kennet, D. Mark, "Did Deregulation Affect Aircraft Engine Maintenance? An Empirical Policy Analysis," *RAND Journal of Economics*, Vol. 24, No. 4, Winter 1993, pp. 542–558.

Kuenne, Robert E., et al., *Warranties in Weapon System Procurement: Theory and Practice*, Westview Press, Boulder, Colorado, 1988.

Kuznik, Frank, "Opening ACTS," *Air & Space*, October/November 1996.

Lane, Polly, "A3XX Launch Postponed," *Seattle Times*, 10 February 1998.

Lane, Polly, "Production Line Is First to Get Stonecipher's Focus," *The Seattle Times*, 27 November 1997.

Lopez, Virginia C., and David H. Vadas, *The U.S. Aerospace Industry in the 1990s: A Global Perspective*, The Aerospace Research Center, Aerospace Industries Association of America, September 1991.

Los Angeles Times, "Airbus to Gauge Demand for Super-Jumbo Jetliner," 9 December 1999.

Mann, Paul, "ITC Study Finds Cost Not Key Element in Aircraft Decisions," *Aviation Week and Space Technology*, Vol. 117, 20 December 1982, p. 30.

Martin, Jim, "Understanding the Terminology and What It Means," *TACTech On-Line Briefs*, Vol. V, No. 3, 1995.

McNaugher, Thomas L., *New Weapons, Old Politics: America's Procurement Muddle*, The Brookings Institution, Washington, D.C., 1989.

McQuiddy, David N. Jr., et al., "Transmit/Receive Module Technology for X-Band Active Array Radar," *Proceedings of the IEEE*, Vol. 79, No. 3, March 1991.

Morris, John, "AlliedSignal Has 12,000 at Work in Europe As Farnborough Helps Boost Firm's Identity," *Aviation Week*'s Farnborough International 98 Show, News Online, 7 September 1998, available on the web at www.awgnet.com/shownews/farnday1.

Muellner, George, Keynote Address, Orlando Air Force Association Symposium, Florida, 16 February 1996.

Myers, Mark, and Fred Bartlett, *Evaluation of Industrial Surface Mount Plastic Encapsulated Microcircuits for Military Avionics Applications*, TRW Avionics Systems Division, San Diego, California, May 1996.

National Aeronautics and Space Administration (NASA), Office of Space Access and Technology, *NASA Spinoff 1995*, NP-217, 1995.

National Economic Council, National Security Council, and Office of Science and Technology Policy, *Second to None: Preserving*

America's Military Advantage through Dual-Use Technology, DTIC #AD-A286779, Washington D.C., 1995.

Neven, Damien, and Paul Seabright, "European Industrial Policy: The Airbus Case," *Cahiers de Recherches Economiques,* No. 9509, Department D'Econometrie et D'Economie Politique, University de Lausanne, June 1995.

Newhouse, John, *The Sporty Game,* Alfred A. Knopf, New York, 1982.

Nordwall, Bruce D., "Honeywell's VIA 2000 Draws on 777 AIMS," *Aviation Week and Space Technology,* 7 August 1995, pp. 43–45.

Northrop Grumman, Electronics & Systems Integration Division, *Northrop Grumman's Technology Reinvestment Project Accomplishments,* July 1996.

Office of Technology Assessment (OTA), U.S. Congress, *Assessing the Potential for Civil-Military Integration: Technologies, Processes, and Practices,* OTA-ISS-611, U.S. Government Printing Office, Washington, D.C., September 1994.

Office of Technology Assessment (OTA), U.S. Congress, *Holding the Edge: Maintaining the Defense Technology Base,* U.S. Government Printing Office, Washington, D.C., April 1989.

Office of the Assistant Secretary of the Air Force for Acquisition (SAF/AQ), U.S. Department of Defense, *Acquisition Reform Success Story: Joint Direct Attack Munition (JDAM),* December 1997.

Office of the Assistant Secretary of the Air Force for Acquisition (SAF/AQ), U.S. Department of Defense, *Acquisition Reform Success Story: Joint Direct Attack Munition (JDAM),* December 1998.

Office of the Assistant Secretary of the Air Force for Acquisition (SAF/AQ), U.S. Department of Defense, *Acquisition Reform Success Story: Wind Corrected Munitions Dispenser (WCMD),* 12 June 1997.

Office of the Deputy Under Secretary of Defense for Acquisition Reform (OUSD/AR), *Cost as an Independent Variable: Stand-Down Acquisition Reform Acceleration Day,* May 1996.

Office of the Deputy Under Secretary of Defense for Acquisition Reform (OUSD/AR), Pilot Program Consulting Group, U.S.

Department of Defense, *Celebrating Success: Forging the Future*, 1997a.

Office of the Deputy Under Secretary of Defense for Acquisition Reform (OUSD/AR), Pilot Program Consulting Group, U.S. Department of Defense, *1997 Compendium of Pilot Program Reports*, 1997b.

Office of the Deputy Under Secretary of Defense for Acquisition Reform (OUSD/AR), U.S. Department of Defense, *Joint Direct Attack Munition (JDAM)*, April 1997.

Office of the Director, Operational Test and Evaluation (ODT&E), U.S. Department of Defense, "Joint Direct Attack Munition (JDAM)," *FY97 Annual Report*, February 1998.

Office of the Under Secretary of Defense for Acquisition and Technology (OUSD/A&T), U.S. Department of Defense, *Dual Use Technology: A Defense Strategy for Affordable, Leading-Edge Technology*, February 1995.

Office of the Under Secretary of Defense for Acquisition and Technology (OUSD/A&T), U.S. Department of Defense, *Overcoming Barriers to the Use of Commercial Integrated Circuit Technology in Defense Systems*, October 1996.

Office of the Under Secretary of Defense for Acquisition and Technology (OUSD/A&T), Acquisition Reform Benchmarking Group, U.S. Department of Defense, *1997 Final Report*, 30 June 1997.

Olney, Ross D., Robert Wragg, Robert W. Schumacher, and Francine H. Landau, *Automotive Collision Warning System*, Automotive Electronics Development, Delco Electronics Corporation, Malibu, California, 1996.

Peck, Merton J., and Frederic M. Scherer, *The Weapons Acquisition Process: An Economic Analysis*, Harvard University, Boston, Massachusetts, 1962.

Perry, William, "Acquisition Reform: A Mandate for Change," Statement from the Secretary of Defense to the U.S. Congress

House Armed Services Committee and Governmental Affairs Committee, 9 February 1994.

Perry, William, "Specifications and Standards: A New Way of Doing Business," Memorandum from the Secretary of Defense, 29 June 1994.

Pint, Ellen M. and Laura H. Baldwin, *Strategic Sourcing: Theory and Evidence from Economics and Business Management*, RAND, MR-865-AF, 1997.

Preston, Colleen A., and Timothy P. Malishenko, "Acquisition Reform," Remarks before the Air Force Association Symposium *Reengineering the Industrial Base*, Dayton, Ohio, 10–11 May 1994, published by the Aerospace Education Foundation.

Proctor, Paul, "Boeing Hones New 550-Seat Transport," *Aviation Week and Space Technology*, 26 April 1999, pp. 39–40.

Red Herring, Interview with Teledesic president Russell Daggatt, Issue 29, March 1996.

Rodgers, Eugene, *Flying High: The Story of Boeing and the Rise of the Jetliner Industry*, Atlantic Monthly Press, New York, 1996.

Rogerson, William P., *An Economic Framework for Analyzing DoD Profit Policy*, RAND, R-3860-PA&E, 1992.

Rolls-Royce Allison Engine Company, Worldwide Customer Support, "Warranty, Guarantee, and Power By The Hour," undated web-page, available at www.allison.com/www/support/pbh-250.html.

Rush, Benjamin C., "Cost as an Independent Variable: Concepts and Risks," *Acquisition Review Quarterly*, Spring 1997.

Sabbagh, Karl, *Twenty-First Century Jet: The Making and Marketing of the Boeing 777*, Scribner, New York, 1996.

Saounatsos, Yorgo E., "Technology Readiness and Development Risks of the New Supersonic Transport," *Journal of Aerospace Engineering*, Vol. 11, No. 3, July 1998, pp. 95–104.

Scherer, Frederic M., *The Weapons Acquisition Process: Economic Incentives*, Harvard University, Boston, Massachusetts, 1964.

Schneider, Joseph, "What Are the Primes Up To? Prospects for Aviation Industry Suppliers," JSA Partners, presentation to the Speednews Aviation Industry Suppliers Conference, 15 March 1998.

Schwendeman, David, "Market-Driven Costing: A Case Study of a Commercial Derivative Airplane Program," presentation for the *Air Force Conference on Cost as an Independent Variable,* Andrews Air Force Base, Maryland, 8–9 July 1997, available on the web at http://www.acq-ref.navy.mil/narsoc/757300/index.html.

Semiconductor Industry Association (SIA), Government Procurement Committee, *Perspective and Recommendations on Diminishing Manufacturing Sources,* 1996.

Sextant Avionique, "International airlines select Sextant Avionique/Smiths Industries new FMS for their Airbus fleets," Webpage, September 1998, available at www.sextant-avionique.com/eng/index.htm.

Shiver, Jube Jr., "Teledesic Set to Fulfill Its Sky-High Goal for Satellites," *Los Angeles Times,* 6 January 1997.

Smith, Giles K., Jeffrey A. Drezner, William C. Martel, J. J. Milanese, W. E. Mooz, and E. C. River, *A Preliminary Perspective on Regulatory Activities and Effects in Weapons Acquisition,* RAND, R-3578-ACQ, March 1988.

Sommer, Geoffrey, Giles K. Smith, John L. Birkler, and James R. Chiesa, *The Global Hawk Unmanned Aerial Vehicle Acquisition Process: A Summary of Phase 1 Experience,* RAND, MR-809-DARPA, 1997.

Stimson, George W., *Introduction to Airborne Radar,* Hughes Aircraft Company, El Segundo, California, 1983.

Stonecipher, Harry, "A New Broom: The Case of Competition in Aircraft Maintenance, Repair, and Overhaul," keynote address to MRO '97 Conference, Dallas, Texas, 15 April 1997.

Sutton, Oliver, "Airbus Snaps at Boeing's Heels in 1997," *Interavia,* January 1998, p. 7.

Sweetman, Bill, "NASA Team Hopes New SST Design Will Reduce Cost, Technology Risks," *Aeroweb*, July 1996, available on the web at www.pollux.com/aeroweb.

Sweetman, Bill, "Coming to a Theatre Near You," *Interavia Business and Technology*, July 1999.

Tang, Christopher S., "Supplier Relationship Map," *International Journal of Logistics: Research and Applications*, Vol. 2, 1999, pp. 39–56.

U.S. Congress, Senate Committee on Armed Services, Acquisition and Technology Subcommittee, *Statement of the Under Secretary of Defense for Acquisition and Technology, The Honorable Paul G. Kaminski, on Defense Acquisition Reform*, 19 March 1997.

U.S. Department of Defense (DoD), Advisory Panel on Streamlining and Codifying Acquisition Laws, *Streamlining Defense Acquisition Laws, Report of the Acquisition Advisory Panel to the United States Congress*, January 1993.

U.S. Department of Defense (DoD), *COSSI Program Description*, 2 February 1999, available on the web at www.acq.osd.mil/es/dut/cossi/ars.html.

U.S. Department of Defense (DoD), *Defense Acquisition Handbook*, 30 June 1998.

U.S. Department of Defense (DoD), *Program Description for the FY1999 Commercial Operations and Support Savings Initiative (COSSI)*, No. 98-17, 31 August 1998.

U.S. General Accounting Office (GAO), *Acquisition Reform: Military-Commercial Pilot Program Offers Benefits But Faces Challenges*, GAO/NSIAD-96-53, Washington, D.C., June 1996.

U.S. General Accounting Office (GAO), *Acquisition Reform: Efforts to Reduce the Cost to Manage and Oversee DoD Contracts*, GAO/NSIAD-96-106, Washington, D.C., April 1996.

U.S. General Accounting Office (GAO), *Airline Deregulation: Changes in Airfares, Service, and Safety at Small, Medium-Sized, and Large*

Communities, GAO, GAO-RCED-96-79, Washington, D.C., April 1996.

U.S. General Accounting Office (GAO), *Precision-Guided Munitions: Acquisition Plans for the Joint Air-to-Surface Standoff Missile*, GAO/NSIAD-96-144, Washington, D.C., June 1996.

van Opstal, Debra, *Road Map for Federal Acquisition (FAR) Reform: Report of the CSIS Working Group*, Center for Strategic and International Studies, Washington, D.C., 1995.

Wilson, J. R., "Suppliers Are Becoming Partners in the New Airframer Paradigm," *Aeroweb*, July 1996, available on the web at www.pollux.com/aeroweb/1996/july1996/supply.html.

Zhang, Ran, "An Economic Analysis of Commercial Jet Aircraft Prices," unpublished Ph.D. dissertation, Boston University, 1996.